AI-Generated Popular Culture

Marcel Danesi

AI-Generated Popular Culture

A Semiotic Perspective

palgrave
macmillan

Marcel Danesi
University of Toronto
Toronto, ON, Canada

ISBN 978-3-031-54751-5 ISBN 978-3-031-54752-2 (eBook)
https://doi.org/10.1007/978-3-031-54752-2

This Palgrave Macmillan imprint is published by the registered company Springer Nature Switzerland AG
The registered company address is: Gewerbestrasse 11, 6330 Cham, Switzerland

Paper in this product is recyclable.

Preface

And nowadays, the idea of AI is really not science fiction anymore. It's just science fact.
—Lisa Joy

Several momentous events occurred in 2023, which signaled that the advent of advanced Artificial Intelligence (AI) systems were radically affecting and changing the ways in which humans did things, including activities that pertained to the creation of original works in the arts. One was a twelve-minute film called *The Frost* (2023), released on June 26, 2023, which gained broad attention from culture critics and the general public alike. Every shot in the film was created by Generative AI (mainly by the DALL-E 2 system), albeit prompted, fine-tuned, and guided by humans at the video creation company Waymark. The film's images of cold icy mountains, a military-style camp, a group of people huddled around a fire, and barking dogs generated contrasting emotions and reactions on the part of viewers. When a man in the video says, with an abrupt tone of voice, "Pass me the tail," a close-up of the strange-looking man standing by the fire gnawing on a pink piece of meat juts out grotesquely at us, with the man's lips masticating the meat in an unnatural, almost freakish, way. For a moment, it appears as if the man is chewing on his own frozen tongue.

The film was just one of countless examples of a new trend that bears profound implications for traditional notions and assumptions about the nature of creativity and the meaning of art in human life. The bizarre video impels us to glean from it some inherent meaning, whatever it might be: Is the source of the meaning in the AI, in the human prompters, or in us the viewers? Is there even an intended meaning in the film? The title of the short film, incidentally, recalls Dante's version of Hell in the *Divine Comedy*. In Canto 32 of the Inferno, Dante puts the worst offenders—the traitors—into a region where they become submerged up to their necks in ice while still in a fiery hell. This film evokes the same Dantesque synesthetic metaphor in a strange new way, with its juxtaposition of the bonfire, the frozen world around it, and a seemingly frozen tongue, creating a jarring sense of something ominous unfolding before us. The characters in the film are in fact debating whether a mysterious sound they hear is a natural phenomenon or something more dangerous. However one interprets the film, it is clear that it heralded the arrival of a machine competitor to humans in the world of filmmaking, with the potential to completely disrupt that world.

A second major event that occurred in 2023 was the labor strike by over 11,000 members of the Writers Guild of America, which began on May 2 and ended on September 27. Among the demands of the writers one stood out—namely that AI must not be allowed to take their place as screenwriters. During the nearly five-month walkout, no issue resonated more among the strikers than the AI one. What was previously a lesser demand of the Guild, had become by 2023 an existential rallying cry. The strike was the first workplace battle between humans and AI, which the humans won, but which may have been a Pyrrhic victory, given the continuing spread of films like *The Frost* after resolution of the strike, and because the new contract left room for studios to train AI systems using preexisting material. The technology itself was clearly not at issue—given that it could be used to assist the writers with summarizing, editing, and carrying out other tiresome tasks. Rather, the issue was the peril of AI replacing them. Two months after the start of the strike, comedian Sarah Silverman sued Meta and OpenAI for copyright infringement, alleging that the companies had trained their chatbots on her books without her consent. Together with the release of *The Frost*, it had become crystal clear that AI had entered the cultural-creative realm, threatening to change it permanently by competing with the humans within it.

As a recurring theme in literary fiction, AI goes back at least to Mary Shelley's *Frankenstein* (1818), followed by Samuel Butler's *Erewhon* (1872). In the former, Shelley explores what a humanoid-looking machine intelligence would look like—a likeable monster, who learns like a contemporary algorithm throughout the novel. In the latter, there is a section titled "The Book of the Machines," in which Butler imagines the possibility of machine consciousness, and the possibility of machines becoming able to replicate themselves. By the early twentieth century, the theme of sentient AI in novels and movies had become commonplace, starting with *Metropolis* (1927), directed by the legendary Fritz Lang, in which a dystopian world was the setting for the terrifying consequences that AI technology entailed. Since then, films involving AI have fallen into several broad thematic areas, including the following two: (1) films revolving around rebellions by robots or computers, such as Stanley Kubrick's 1968 film, *2001: A Space Odyssey*, with its murderous onboard computer; and contrariwise (2) films dealing with the potentially more positive aspects of AI, as can be seen with the R2-D2 robot in George Lucas' 1977 *Star Wars* and the eponymous robot in the 2008 *WALL-E* movie. However, the warnings and praises of intelligent AI have now moved beyond the sci-fi world, with AI itself having entered that very world, with its own films, such as *The Frost*.

Generative AI raises obvious questions about creativity and its role in shaping pop culture, which has been a mainstream one for at least a century across many urban societies. AI is now used widely in film, radio, and television for content creation. In music, it is employed for producing songs on demand. In video games, it has the ability to create more realistic and engaging adventure situations. AI-generated novels and poetry are now esteemed for their "originality," and are even awarded prizes. But, to reiterate, is the originality in the machine, the programmer, the audience, or somewhere else? What is the nature of the semiotic dynamic (the meaning-making impulse that humans imprint in their representations) in an AI-produced work? The future has arrived, and it is not different from what the sci-fi novels and movies had predicted.

As far as I can tell, no comprehensive discussion exists addressing such questions. Such a treatment is, I believe, critical at this moment. The particular slant I have taken toward this topic is that of semiotic analysis, defined broadly as the science of meaning, because the real problem with the AI trend concerns meaning—what it is, how we grasp it, why we need it. No technical knowledge on the part of the reader is assumed—I will explain any technical term concretely at specific points in a discussion, if it comes up. One key question that a semiotic perspective raises, and which will be addressed throughout this book, is the following one: How do AI-generated texts affect what the philosopher-semiotician Charles Peirce called the *interpretant*, which can be defined simply, for the present purposes, as a shifting mode of understanding the meaning or function of some text in terms of the time, place, situation, and individual involved in creating it?

I should mention that in the period in which I wrote this book, Generative AI systems were said by experts to be in "early stages," although it was hard for me to discern in what sense the word "early" was used, given the highly sophisticated forms produced by the AI tools I examined at the time. The other common phrase employed was "we are only at the beginning." If so, then we are truly at a crossroads era that will have far-reaching consequences. Of course, technology is always evolving. Nevertheless, it is in the "earliest" stages that the potential of a technology is best assessed.

Toronto, Canada
2023

Marcel Danesi

Abbreviations

Note that abbreviations will be used throughout. For example, "AI music" is used typically to stand for "AI-generated music," "AI novels" for "AI-generated novels," and so on.

Contents

1	**AI, Popular Culture, Semiotics**	1
	Prologue	1
	Generative AI	3
	AI in Science Fiction	5
	AI-Generated Popular Culture	8
	A Semiotic View	13
	Hyperreality	16
	Epilogue	18
	References	19
2	**AI-Generated Literature**	23
	Prologue	23
	The Role of Fiction	25
	The Author and the Reader	28
	AI Novels	31
	AI Poetry	34
	AI Comics	37
	AI and History	39
	Epilogue	41
	References	42

3 **AI-Generated Cinema** 45
Prologue 45
Movies and Popular Culture 47
AI Filmmaking 49
AI Movies 53
Whither Cinema? 57
AI and the Theory of Mind 59
Epilogue 61
References 63

4 **AI-Generated Music** 67
Prologue 67
Music and Popular Culture 69
AI Music 71
AI and Classical Music 75
Android Performers 80
Epilogue 82
References 83

5 **AI in the Mass Media** 87
Prologue 87
AI Radio 90
AI Podcasts 93
AI Television 95
Mediated Social Codes 97
Epilogue 100
References 103

6 **AI-Generated Art** 105
Prologue 105
Pop Art 107
AI Art 110
The Aura 117
Algorithmic Art Criticism 119
Semiosis 121
Epilogue 123
References 124

7	**AI in Marketing and Advertising**	127
	Prologue	127
	Advertising and Pop Culture	129
	AI Ads	132
	Hyperreal Marketing	136
	The Frankfurt School Redux	138
	Epilogue	140
	References	141
8	**AI and Gaming Culture**	143
	Prologue	143
	Video Games and Popular Culture	144
	AI Video Games	148
	The Simulation Hypothesis	150
	AI-Generated Narrative Content	152
	Epilogue	154
	References	156
9	**Simulacrum Culture**	159
	Prologue	159
	Simulacrum Theory	161
	Virtual Reality	163
	Deepfake Culture	166
	The Metaverse and the Narraverse	169
	Epilogue	171
	References	173
10	**AI and the Future of Pop Culture**	175
	Prologue	175
	Why Pop Culture?	177
	A Theory of AI-Generated Pop Culture?	180
	AI Creativity	182
	Where's the Beef?	184
	Six Characters in Search of an Author	185
	Epilogue	187
	References	189
Index		193

List of Figures

Fig. 1.1 The McLuhanian Epochs 12

Fig. 2.1 Queneau's *Cent mille milliards de poèmes* (1961) (*Source* Wikimedia Commons, https://commons.wikimedia.org/wiki/File:Queneau_sonnets.jpg. Creative Commons Attribution 2.0 Generic: "This image was originally posted to Flickr by Thomas Guest. It was reviewed on 19 August 2023 by FlickreviewR 2 and was confirmed to be licensed under the terms of the cc-by-2.0") 32

Fig. 4.1 Piano-Playing Android, 2021 (Adobe Stock: https://stock.adobe.com/bg/images/ai-android-robot-playing-piano-generative-ai-illustration/589926432) 81

Fig. 6.1 AI-Portrait, Emilie Arnoux (Wikimedia Commons: Creative Commons Attribution-Share Alike 4.0 International) 109

Fig. 6.2 AI-Portrait, *Edmond de Bellamy* (Wikimedia Commons, https://commons.wikimedia.org/wiki/File:Dmond_de_Belamy.png) (Attribution: "This file is in the public domain because it is the work of a computer algorithm or artificial intelligence and does not contain sufficient human authorship to support a copyright claim") 111

Fig. 6.3 *Théâtre d'Opéra Spatial*, Midjourney, 2022 (Wikimedia Commons, https://commons.wikimedia.org/wiki/File:Théâtre_D'opéra_Spatial.png) (Attribution: "This file is in the public domain because it is the work of a computer algorithm or artificial intelligence and does not contain sufficient human authorship to support a copyright claim") 112
Fig. 6.4 Painting made by Stable Diffusion (Creative Commons CC0 1.0 Universal Public Domain Dedication. https://en.wikipedia.org/wiki/File:An_astronaut_riding_a_horse_%28Picasso%29_2022-08-28.png) 114
Fig. 6.5 DALL-E's Generation of A *Painting of a Hand in the Style of Van Gogh* (Wikimedia Commons, https://commons.wikimedia.org/wiki/File:A_painting_of_a_hand_in_the_style_of_Van_Gogh_-_DALL-E.png) (Attribution: "This file is in the public domain because it is the work of a computer algorithm or artificial intelligence and does not contain sufficient human authorship to support a copyright claim") 115
Fig. 7.1 AI-Created ad for a fictional pizza (from Twitter, @Pizza_Later) 134
Fig. 7.2 Deepfake Will Smith video (Screenshot from YouTube) 135
Fig. 9.1 Deepfake *Mona Lisa* (Midjourney, Wikimedia Commons) (Info: A rendition of the Mona Lisa by Midjourney AI. Created with the prompt "Oil painting of the Mona Lisa by Leonardo DaVinci": https://en.wikipedia.org/wiki/File:Mona_Lisa_by_Midjourney_AI.jpg: (Attribution: "*This file is in the public domain because it is the work of a computer algorithm or artificial intelligence and does not contain sufficient human authorship to support a copyright claim.*") 167

CHAPTER 1

AI, Popular Culture, Semiotics

The thing that's going to make artificial intelligence so powerful is its ability to learn, and the way AI learns is to look at human culture.
—Dan Brown

Prologue

The pioneer American computer scientist Arthur Samuel designed one of the first rudimentary computer programs in 1952 capable of adapting independently to changing data, using the term "machine learning" to describe the process involved (Samuel 1959). Called the Samuel Checkers-Playing Program, it was among the world's first successful self-teaching programs, with the ability to learn from the "experience" of playing different games of chess. Decades later, IBM developed a chess-playing computer named Deep Blue that was capable of analyzing millions of chess positions per second and to learn strategically from the evolving configurations of the chess pieces on the board, allowing it to optimize its moves. Although it lost its first competition in 1997 to chess champion Garry Kasparov, in a rematch the computer defeated him soundly. Right after that historic game, Deep Blue was disassembled for obscure reasons. The event showed the entire world, nonetheless, that Artificial Intelligence (AI), bolstered by machine learning algorithms, had become truly "intelligent."

M. Danesi, *AI-Generated Popular Culture*,
https://doi.org/10.1007/978-3-031-54752-2_1

In 2011, self-learning AI entered pop culture in a sensationalistic way when the TV quiz show *Jeopardy* featured two human champions competing against IBM's Watson, an AI system designed on purpose for the event. Watson won the match by a large margin. It was the first time that an intelligent machine came to the general public's awareness broadly, becoming a sort of "celebrity" for a while. Lastly, in 2016, another AI system, named AlphaGo, beat the world's Go champion at the time, Lee Sedol, with a creative move that was previously unknown, surprising Go experts. Yet another algorithmic celebrity was born, as over 200 million people worldwide watched the match.

Popularized events such as these have made it saliently obvious that AI is now embedded among us, a fact that bears many implications for the future course of humanity itself. If an AI system can be devised to come up with a truly intelligent move in the game of Go, previously unbeknownst to humans, then a host of related questions arise: Can AI do virtually anything that humans can? Can it generate cultural artifacts, such as songs, novels, paintings, and the like? A positive answer seems to have come forth putatively to such questions. AI-generated paintings now fetch millions of dollars at auctions, computer-generated songs make it to the top of the charts, and films such as *The Frost* (preface) gain acclaim from critics and audiences alike. The lines between human and machine creativity seem to have become rather blurry, to the point of being almost nonexistent. Technology has always been a key factor in spurring on new cultural trends and artistic genres since at least the 1920s, with the advent of sound recording, filmmaking, and radio technologies (Danesi 2023). Together, the technology-pop-culture interlinkage has had an undeniable impact on lifestyles and aesthetic values. So, it should come as no surprise to witness that AI, as a new technology, is now a force changing the world, akin to technologies of the past. Its ability to *create* novel cultural artifacts in a quasi-autonomous fashion, however, is unique in human history, a feat that has far-reaching implications. As Max Roser cogently put it in 2022: "Despite their brief history, computers and AI have fundamentally changed what we see, what we know, and what we do. Little is as important for the future of the world, and our own lives, as how this history continues."

Some of the repercussions that this mindboggling paradigm shift in human evolution bears will be deliberated initially in this opening chapter, against the theoretical backdrop of semiotics as a science of meaning-making. The idea that we evolve cognitively and culturally through technology was formulated concretely and insightfully by the late Canadian communications theorist Marshall McLuhan already in the 1950s and 1960s, at around the same time that Samuel and other computer scientists were experimenting with machine learning. In several key works (McLuhan 1951, 1962, 1964), McLuhan put forth his theory that technologies are designed to be extensions or amplifications of human faculties. So, for example, tools such as an ax, a wheel, and a telescope have allowed humans to extend specific biological capacities—an ax extends the ability of the human hand to break wood; the wheel of the foot to cover greater distances; and the telescope of the eye to see further. In this framework, it appears logical to assume that Generative AI has extended the ability to create art and culture. Or has it?

AI-generated, drawings, books, movies, videos, and songs are now practically indistinguishable from those created by humans. AI algorithms can analyze vast amounts of data and generate novel works, such as new songs, based on that analysis, thus potentially breaking the historical flow of a new musical trend as building on a previous one *intentionally*. Consider, as two early cases-in-point, the AI platforms Amper Music and AIVA (Artificial Intelligence Virtual Artist). At the time of writing this book, the former had actually closed down. It allowed users to get customized music, by prompting the platform to compose a work with a desired mood, genre, melodic style, and tempo. AIVA, which was still available at the time of writing this book, has the ability to create original classical music. It has actually been used to generate music for films, advertisements, and concerts. It has also released full-length albums which have garnered broad critical praise. Clearly, we are at an epic crossroads in human history, impelling us to reconsider our traditional views of creativity, culture, the arts, and human uniqueness. AI is poised to transform the world, radically changing the ways we consume and interact with artistic content.

Generative AI

While AI-generated music has the potential to push the boundaries of traditional music-making, fundamental questions arise in this area, similar to those raised by AI-generated films such as *The Frost*: Does it have emotional depth or is it just "music-by-design" that cannot possibly evoke the same type of aesthetic responses as a human-composed piece of music, such as a symphony by Beethoven? Or is AI music, like the AI systems with the ability to play checkers, chess, and Go inventively, changing our views of what is creativity?

Discussing a Generative AI YouTube video titled *Harry Potter by Balenciaga*, Kyle Chayka (2023) has discerningly observed that online audiences are starting to make little or no distinction between what is created by humans and what is generated instead by AI, as long as it is entertaining or interesting in itself. Against a background soundtrack of electronic music, the simulated Harry Potter characters in the video appear to blink, nod, and utter lines adapted from the films remixed with references to fashion items. As Chayka notes, the amazing thing is that: "In three weeks, it has received almost five million views; a sequel, released less than a week ago, has netted more than a million and a half." Seemingly, the viewers were indifferent to the fact that the video was a deepfake reimagination of the cast of Harry Potter, imitating a 1980s-style commercial for the Spanish luxury fashion house Balenciaga. By portraying the Harry Potter cast of characters as Balenciaga models, the video appeared to be poking fun at the self-important imagery associated with the brand. The comments about the video on social media indicated that people actually had differing views of the video—some critiquing it as a desecration of the Harry Potter saga, others seeing it instead as an updating of the classic characters as fashion models, and others interpreting it as a parody of high fashion advertising and lifestyles. The bizarre video begins with Hagrid telling a fashionably-dressed Harry Potter "You're a Balenciaga, Harry," imitatively recalling the well-known phrase from the original saga, "Harry, yer a wizard." The video continues with a montage showing Harry and other characters wearing stylish attire. At one point, the character representing Dumbledore then says, "To the well-organized mind, Balenciaga is but the next great adventure." The video even ends with Harry casting an Unforgivable Curse: "Avada Balenciaga." The key question that the video raises is articulated insightfully by Rebecca Jennings (2023): "What's interesting…is the question of just how long we, as a society, have before AI-powered video becomes most of what we think of as visual entertainment."

Generative AI is clearly capable of producing human-like musical and video texts. It achieves this feat by learning from the content, patterns, and structure of its input training data, from which it can generate new texts with similar characteristics. Is this a form of creativity that is similar to the human type, or is it different, given that humans also rely on previous texts to create new ones? Human creativeness results from the interaction between the body, the sensory system, perceptual mechanisms, and the imagination, which cannot be separated in any act of true human intellect (Uexküll 1909). So, are intelligent machines creative in the human sense, without a body and a sensory apparatus? As Maturana and Varela (1973: 16) have observed, creativity "takes place in the dynamics of the autonomy proper to living systems," while, as McGann (2000: 358) remarks, AI creativity is the "product of [systems] functioning differently from themselves." The human brain has evolved through embodiment in a physical system—the human body itself—that interprets information in the world and then acts accordingly on it. Most of the currently used AI systems lack any direct connection to the physical world. AI agents such as robots and virtual assistants can of course interact with and learn from the physical environment with sensors (cameras, pressure sensors, accelerometers, etc.). But the embodiment is artificial, not biological, and hence different from human embodiment.

Of course, one could argue that any comparison between humans and machines is a moot one, if the latter can pass the so-called Turing Test, whereby humans themselves cannot tell whether some artifact such as a video or a song is produced by human or artificial intelligence. It is also beside the point to envision machines as possessing self-awareness as do humans. This whole area of research is beyond the scope of the present work. What is of direct interest here is how humans themselves interpret AI creations, raising the key question of traditional hermeneutics: Where is the meaning of a text and who determines what that meaning is? This question is a primary thematic undercurrent guiding the discussion in the remainder of this book.

AI in Science Fiction

Portraying and dramatizing the implications of intelligent sentient machines has been a major interest of science fiction writers, starting (as discussed in the preface) with Mary Shelley's *Frankenstein* and Samuel Butler's *Erewhon*, considered to be prototypes for the sci-fi genre that emerged later in the nineteenth century as a distinct literary category. The portrayals involve an array of contrasting visions, from dystopian to utopian ones, where robots either destroy human civilization or else help humans build a better one (Osawa et al. 2022).

A common theme in the genre is the "Frankenstein complex," a term coined by sci-fi writer Isaac Asimov (1950: 106), whereby a robot turns on its creator. Examples of this complex in cinema include the 1982 film *Blade Runner*, in which the replicants gouge out the eyes and crush the skull of their human creator, and the 2015 film, *Ex Machina*, where the intelligent entity Ava stabs her creator to death. Another major theme, starting with Karel Čapek's 1920 play *R.U.R.*, is that of AI rebellion, in which robots are depicted as usurping control over humans, forcing people into submission and slavery. Alternatively, in other stories, humanity is actually saved from itself by robots, which have been programmed to be of service to humans, as for instance in Isaac Asimov's 1942 short story *Runaround*, where we find his famous "Three Laws of Robotics": (1) A robot may not injure a human being or, through inaction, allow a human being to come to harm; (2) A robot must obey the orders given to it by human beings except where such orders would conflict with the First Law; (3) A robot must protect its own existence as long as such protection does not conflict with the First or Second Law. The theme of simulated reality has also become a common one in more recent sci-fi, starting with the 1999 film *The Matrix*, which depicts a world where humans are trapped within a computer simulation. The film has come to constitute a fictional framework for the so-called simulation hypothesis, to be discussed subsequently, which claims that what we perceive as reality is actually a simulation, overseen by an advanced AI.

One of the first mentions of robots (called automata) is in Homer's *Iliad,* in which the god of metalwork, Hephaestus, created an army of robot servants. By the time of Mary Shelley's *Frankenstein* in 1818, the idea that robots could be brought to life fascinated and horrified the popular imagination. The term *robot* comes from the Czech word *robota*, meaning "drudgery," used for the first time with its modern designation by Čapek in *R.U.R.* (above).

Whatever the thematic slant of the sci-fi novels or movies, the kind of scenarios depicted in them have a common subtext—technology has become increasingly intertwined with human progress to the point that it could easily run human affairs with little or no human input. American media critic Neil Postman (1992) put forward a warning about this vision, calling a world dominated by technology, a "technopoly," which he defined as a society which believes that "the primary, if not the only, goal of human labor and thought is efficiency, that technical calculation is in all respects superior to human judgment." Films dealing with technopolist worlds, from *Westworld* (1973), *Demon Seed* (1977), and *The Terminator* (1984), to *Ex Machina* (2014), *Brian and Charles*

(2022), and *The Creator* (2023), among many others. The movie that appears to best portray this world, in which a human–machine partnership is inevitable, is Ridley Scott's *Blade Runner* (1982). The robot in this movie is called a replicant, a bioengineered humanoid that is physically indistinguishable from a human, but possessing superhuman strength and intelligence. A subtext of the film is, arguably, that human social evolution has been altered radically by our excessive dependence on technology. Based on the 1968 story *Do Androids Dream of Electric Sheep?* by Philip K. Dick, the movie asks the fundamental questions of philosophy: What is a human being? What is real? Is there any purpose to existence? These questions are evoked, however, through the motives and actions of the replicants, whose struggle to survive extinction mirrors the human struggle to do so.

To assess the social interpretations and impacts of such movies, Hudson et al. (2021) interviewed AI specialists, technologists, science fiction writers, and policy experts, aiming to identify key ideas in them that were relevant to AI technologies. One of the themes that emerged from the interviews concerned an ever-evolving posthuman world, in which humans will have to become accustomed to living in consort with machines, as in *Blade Runner*. The researchers concluded that the depictions in the movies and actual AI research, however, may not always match, given that it is dubious (in their estimation) that "AI that is equivalent to a human being will ever be attainable or even conceptually useful."

The starting point for debating the nature of a human–robot partnership is Mary Shelley's *Frankenstein*. As Byrd and Paquette (2023) note, "this two-century-year-old tale provides a blueprint for understanding the conflict among humans and machines and provides a roadmap for harmonization in the past, present, and future." As mentioned, Samuel Butler broached the notion of machines with superhuman intelligence shortly after Shelley's avant-garde novel, in *Erewhon*, in which he raised the question of the possibility of consciousness among self-replicating machines and the danger that the machines might supplant humans as the dominant species. George Eliot also explored this possibility a few years later in her last work, *Impressions of Theophrastus Such* (1879). In the chapter titled "Shadows of the Coming Race," Eliot depicts an argument between the characters Theophrastus and Trost about the future use of thinking machines. Theophrastus claims that humans will be supplanted by thinking machines, while Trost counters by maintaining that the machines will improve humanity, by making menial and repetitive labor a thing of the past, freeing humans to contemplate higher things. To cite Helen Small (2020), the contemporary relevance of Eliot's novel was not "its imagining of future machine intelligence," but rather "its prediction of how the cultural debate has developed around AI and AI's consequences for humanity."

AI is now a fact of everyday life, and it is moving into all domains that were once controlled exclusively by humans, reshaping how we live, think, and act, much like the science fiction works predicted. As Isaac Asimov so aptly put it: "Science fiction always bases its future visions on changes in the levels of science and technology. And the reason for that consistency is simply that—in reality—all other changes throughout history have been irrelevant and trivial" (cited in Stone 1980).

AI-Generated Popular Culture

The ability of AI to generate cultural artifacts virtually autonomously, including novels, poetry, songs, games, videos, advertising, and even artificial brands, is a truly remarkable feat of human technology. But how do the AI artifacts compare with those produced by human creativeness and originality? Does the AI pass a kind of creative Turing Test or, more concretely, a Voight-Kampff Test? The latter is a fictional interrogation tool, originating in Philip Dick's story, *Do Androids Dream of Electric Sheep* and the movie *Blade Runner*, used by human blade runners to determine whether an individual is a replicant or not. It is an empathy test, which detects small changes in facial expression (and especially the eyes). It is described in the film's press kit as follows (cited in Edwards 2023):

> A very advanced form of lie detector that measures contractions of the iris muscle and the presence of invisible airborne particles emitted from the body. The bellows were designed for the latter function and give the machine the menacing air of a sinister insect. The VK is used primarily by Blade Runners to determine if a suspect is truly human by measuring the degree of his empathic response through carefully worded questions and statements.

Interestingly, Steven Spielberg's 2001 film, *A.I. Artificial Intelligence*, suggests that rather than the machine itself passing the Voight-Kampff

Test, it is humans who react to the machine as if it were empathic. It thus locates the test in human interpreters. This can be called a "Voight-Kampff Effect," rather than a "Test" (designed for machines). In the film, the android called David has been programmed to display the emotion of love. Because he is a sophisticated anthropomorphic android our tendency is to believe that he does actually experience this emotion, even though in reality, being a disembodied system, David is simply acting out the forms of the emotion according to his coding. In effect, David activates a Voight-Kampff Effect in audiences (Gomes 2017).

A central feature of human art and culture is its "time-binding" nature, as the Polish-American scholar, Alfred Korzybski (1921) called it. This refers to the distinctively human attribute of preserving memories and records of experiences for the use of subsequent generations. In other words, artworks are products of continuous historical processes, connecting them across time. As Korzybski (1921) defined it, time-binding is "the capacity of an individual or a generation to begin where the former left off." If the process is interrupted or destroyed, it would signal the end of culture as a holistic entity and the need to virtually start over. Even works that break away from the past are nonetheless inspired by it, often constituting an act of rebellion or liberation. Abstract art, for instance, can be understood in this way, since it reflected a need initially to escape from the past and even from reality, as Vered Aviv (2014) claims, allowing the brain to "flow within its inner states, create new emotional and cognitive associations, and activate brain-states that are otherwise harder to access."

Pop culture, like all other forms of culture, is a time-binding phenomenon, having a basis in previous creative artifacts that form a hereditary lineage. It is highly appealing for this very reason, since it projects the past into the present via its established artistic objects, stimulating new forms of creativity. A simple example would be the echoes of early jazz in subsequent musical genres such as rock and roll and hip-hop. These would not have come about in the first place without jazz. In the case of the birth of hip-hop in the late 1970s, the influence of jazz can be seen concretely in the massive number of samples used from the jazz genre guiding hip-hop's development. From rhythms to turns of musical phrase, jazz music continues to reverberate in the hip-hop genre, no matter how distant the echoes.

According to the time-binding principle, innovations in human culture and art do not crop up in a vacuum, but rather are inspired by the flow of aesthetic forms that accrue over time. Without this, we would be cut off from the art of the past, making present art-making difficult to realize. AI-generated art, on the other hand, is not bound by this principle, since it generates its products via a combination or permutation process of the pre-existing forms on which it was trained. However, the paradox is that if an AI-generated artistic text, no matter how it was put together, evokes the Voight-Kampff Effect in us, then all one can say is that it is *ipso facto* a form of art. The Voight-Kampff Effect is now everywhere, inducing audiences to even envision fictional (avatar) rappers as legitimate artists in their own (weird) right. One well-known example is that of FN Meka, described as a "virtual rapper," who was generated in 2019 by an AI trained on thousands of data points compiled from video games and social media. Incredibly, the avatar rapper and his AI-generated songs attracted over 10 million followers and over a billion views on TikTok. He was then signed by Capitol Records in 2022, becoming the first "AI-Generated rapper" to be signed by a major label. However, due to controversy over criticism about the appearance of FN Meka as a stereotyped male African American, Capitol dropped the avatar rapper from its label a little later.

Now, at one level, AI is just a current technology for creating popular artworks, and one cannot underestimate the role of technology in the origins and spread of pop culture itself. The rise of music as a mass art, for instance, was made possible by the advent of recording and radio broadcasting technologies in the early part of the twentieth century, making the music available to large audiences. The spread has become even broader with television and Internet technologies, which have been diffusing popular forms of music throughout the globe. The difference between AI and technologies of the past, however, is that the latter were media for both the creation of new forms of art by humans (as in film-making) and for transmitting the art to large audiences, whereas AI is often itself the creator and transmitter, albeit with human help at the present time.

Let us consider, as an example, OpenAI's "MuseNet," called anthropomorphically an "AI composer." The algorithm is trained on a massive dataset of pre-existing musical compositions, from which it is able to generate works similar to those of the human composers, such as symphonies, concertos, and piano solos. Now, the question becomes: Does such music garner the same kinds of emotional responses as do the human-generated works? There are sufficient studies to indicate that it does not have the same emotional pull of music created by human composers like a Bach or a Beethoven (e.g., Novelli and Proksch 2022). But this is not always the case. AI's 2023 deepfake track "Heart On My Sleeve," did seem to have this type of pull, perhaps because of its uncanny resemblance to what seemed like a musical cooperation between the music artists Drake and The Weeknd. The AI song, in fact, garnered millions of views on social media networks, and generated many positive comments as to its emotional effects. A large number of listeners were apparently unable to detect any difference between the fake song and what the artists might themselves have created. Others seemed simply to not care, perhaps because the AI voice filters used to imitate Drake and The Weeknd were indistinguishable from the real voices—after all, imitation is the greatest form of flattery. However, as Shamira Ibrahim (2023) remarks, an imitation such as this one cannot possibly match the human original:

> Fans who are enthusiastic about the future of generative AI models in music are misguided at best. As a musical genre and cultural hub, hip-hop is eternally innovative, having made great strides on autonomous and creative uses of technology to advance the landscape, from quick mix breakbeats on turntables to samples on an MPC drum machine, to digital recordings. It takes only one listen to "Heart on My Sleeve" to hear what's lost—the magic that comes with organic performances with the human voice; the patented ad-libs that become an artist's signature, as well as a chart for a record's emotional tempo; the flat notes that aren't pitch corrected.

Whatever the truth of the matter, the emergence of AI-generated music sheds light on the remarkable ability of AI systems to produce lyrics, melodies, rhythms, and harmonies that capture the characteristic traits of specific artists with a high degree of accuracy and realism. Even if it is known that the music was generated by AI, it nevertheless brings about the Voight-Kampff Effect in many people. AI-generated music is thus leading to a redefinition of the roles of the composer, artist, and the audience. The meaning of a musical work may well be inherent in the design of the work itself, no matter who or what created it, and how it was created. As Tomas Chamorro-Premuzic (2023) aptly points out, "there is no shortage of examples to illustrate the creative capabilities of machines, as measured by human standards." As he goes on to remark, AI can improvise works that are almost identical to those of a Miles Davis "fooling 90% of Jazz experts…and [even] complete Schubert's renowned Unfinished Symphony."

To cite McLuhan again, it is important to emphasize that the current AI trend in artistic creativity is not unexpected, since societies evolve in accordance with changes in technology, not independently of them, bringing about the different epochs of history. The earliest epoch was based on the technology of the spoken word, as McLuhan (1962) called it, whereby knowledge was passed on orally to subsequent generations. This social system evolved textual forms (such as myths and legends) based on oral semiosis. The transition to a literate epoch came about when writing systems emerged allowing people to record knowledge in a more permanent and transportable form—via hand-writing. As writing expanded, humanity eventually entered into what McLuhan called the "manuscript epoch"—a period of history in which manuscripts dominated and guided social activities. With the advent of moveable type technology in the late 1400s, the manuscript epoch was transformed into a broader "Print Age." The mass production capacities of the new technology guaranteed that print culture would envelop social systems across the globe, since print materials could be sent across the world, regardless of political borders. By the early twentieth century, the print-based world was being supplanted more and more by electronic communications technologies. Because electronic media have the capacity to transcend the limitations of place and time of previous technologies, a new world order was emerging with the capacity to unite the world into what McLuhan called an "electronic global village."

To this McLuhanian paradigm we can add the current AI Age, which includes digital technologies—McLuhan passed away at the threshold of the digital era. The following diagram can be used to summarize the epochs, noting that each one is a derivative of a previous era (Fig. 1.1):

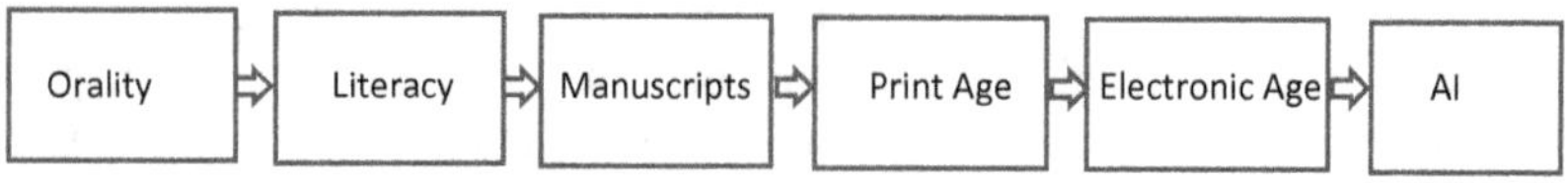

Fig. 1.1 The McLuhanian Epochs

A Semiotic View

As a science of meaning, it can be suggested that semiotics is an appropriate tool for discussing the nature of artistic and cultural artifacts. There are five key notions that are relevant in this regard—text, code, representation, interpretant, and semiosis.

The *text* is a structure made up of combinations of signs, such as words and images, which form a coherent (meaning-bearing) arrangement that we identify as a story, a drawing, a song, and so on. As such, it "stands for" something in the world or even just in the human imagination itself. Constructing a text requires knowledge of cultural *codes*, such as language, musical styles, and the like, which allow for the conversion of ideas or feelings into specific textual forms. *Representation* refers to how the text depicts or portrays what it stands for, and the *interpretant* is the disposition of an interpreter to respond to the text, including the person's subjective perception of, and affective reactions to, the representation itself. It is the effect that a semiotic structure (such as a text) has on someone who reads, views, or comprehends it in some way. This is a key notion, traced to the work of Charles Pierce (1931, 1958), who divided it into three phases: (1) the immediate interpretant, which refers to the interpretability of the text itself; that is to say, a text must lend itself to some interpretation, otherwise it would be perceived as a randomized pastiche of meaningless forms; (2) the dynamical interpretant which is the actual meaning assigned to, or extracted from, the text; and (3) the final interpretant which is the ideal interpretation of that text, based on its intended purpose. Note that the interpretant also guides the creators of a text; it is a bidirectional property of semiosis. Finally, *semiosis* is the innate ability to produce and comprehend sign structures, such as texts, as saying something, or intending to say something, about the world. These definitions have been tailored specifically for the purposes of the present discussion, so that the terms can be used in a general, rather than a technical, way.

Now, the question becomes, does an AI-generated form, such as the song "Heart On My Sleeve," manifest such semiotic properties? It can certainly be identified as a musical text that reveals a specific melodic, harmonic, and rhythmic structure that is characteristic of hip-hop. The code used in this case is an algorithmic one, which is nonetheless trained to access human-made songs, and then rearrange their elements into a novel hip-hop song. In terms of representation, "Heart On My Sleeve" can be seen to represent the music of two human artists in simulated style; so the interpretant is shaped by the same pattern of meanings assigned to the usual music of the two artists, including its lyrical, melodic, harmonic, and rhythmic qualities. Now, the question of semiosis is much harder to address in this case, without hypothesizing what has been called "machine semiosis" (Nöth 2002). But then this would imply that the AI had its own intent for creating the song, while the intent was actually that of the human prompter who, nonetheless, could not predict what the machine's final output would have been. Clearly, the deepfake song poses a profound semiotic dilemma, which will be discussed throughout this book.

The time-binding principle also comes into the overall semiotic picture. The songs of Drake and The Weeknd were devised initially within a specific temporal and social context, but have since evolved into novel creations via the creativity of the artists. The songs that are produced by AI, on the other hand, are ad hoc artificial creations that simulate human texts and their time-bound representations. The machine has no sense of the historical flow of representation and no intent on its own to produce musical art. So, "Heart On My Sleeve" bears no discernible interpretant in itself; it is imagined by human listeners, who assign their own time-bound interpretants to it. Of course, should a future intelligent machine somehow be able to produce new music on its own, with no human prompting, and outside of its algorithmic training, then we would be faced with a truly profound semiotic dilemma: What would it represent? Would it truly reveal a form of machine semiosis to which we might not have access, since we do not share the same algorithmic code?

At this point a key notion, which the Italian philosopher Giambattista Vico (1725) called the *fantasia*, can be enlisted perhaps to help shed some light on the plausibility of machine semiosis. This is a blend of imagination and fantasy that humans employ to understand the world and then to remake it on their own terms, via their imaginative products such as language, visual art, poetry, myths, stories, and so on. Does AI possess a Vichian-type *fantasia*? If so, it would mean that the machine would want to understand the world and its own inner thoughts, which it would then use to remake the world on its own terms. Apart from what that artificial world might look like to human brains, what kind of algorithmic codes, designed by machines themselves, would be necessary? Can human-independent artificial minds indulge in the same kinds of imaginative thoughts produced by the *fantasia*? In sci-fi this is of course a reality, as discussed above. But the imaginative thoughts of an AI would be interpreted anthropomorphically, not in any new conceptual way that would assign thought independence to the machines. Current programs devised by machines themselves are realized by combining existing lines of code taken from other software, not *ex nihilo* as in the case of humans devising novel creations. Humans create for a purpose or intent through the *fantasia*; and thus understand their own creations on their own terms. Machines do not have any identifiable intent in creating something anew and on their own terms. If AI truly becomes autonomous, then it will have to exhibit features of the *fantasia*.

The term *text* requires some further commentary here, given that AI is a generator of texts to which human interpreters assign meanings. Whether it is a work of literature (such as novel), a drawing, a street sign, a scientific diagram, a building design, a style of clothing, a love letter, a conversation with a psychotherapist, etc., a text is designed to relay some message and evoke an interpretation based in a context. The word *text* comes from Old French *texte*, which in turn is derived from the Latin verb *texere* ("to weave together"). It was coined originally to describe how words were used to create a written document—that is, how they were "woven together" to deliver an intended message. A text is not created in a social vacuum—it is meaningless until it is "read" or interpreted by someone in a specific *context* (a communicative event, an artistic performance, and so on), an idea that goes back to Aristotle (350 BCE). For the Greek philosopher, the goal of texts was to explain the purpose, function, and existential meanings of complex things, from mathematics and physics to language and narratives.

It was in the medieval period that fictional texts gained a foothold in the popular imagination, starting arguably with the works of Boccaccio and Chaucer (to be discussed in Chapter 9). By the nineteenth century, the broad spread of creative literary texts, such as novels, modern humans may have developed a state of mind which sees little or no difference between fiction and reality, leading to what Jean Baudrillard (1981) called a "simulacrum effect," which is more than just a suspension of disbelief, or the willing avoidance in perceiving fiction as unreal in order to believe it for the sake of enjoying the narrative, but rather an immersion into the fictional text with little desire to escape from it back to reality. Now, when fictional representations occur on electronic screens repeatedly the simulacrum effect becomes embedded so deeply in the subconscious that it induces the sense that there is no difference between what occurs on either side of a screen, that is between reality and fiction.

In fact, what is fictional is perceived as more real than real, a phenomenon that Baudrillard called an immersion in *hyperreality*.

This term is particularly relevant to the phenomenon of AI-generated art because it speaks directly to the larger social and aesthetic-critical concerns that such art raises. Through the, simulacrum, a textual representation of reality on the screen appears as a self-contained one that is constantly changing, and perhaps it is this lens that induces many of us to perceive an artificial work, such as an AI song, as just *that*, a song, even if it belongs to the realm of the hyperreal.

Hyperreality

The notion of hyperreality requires further commentary at this point, given its relevance to the purposes of this book. It is significant to note that Baudrillard's notion was adopted by the 1999 movie *The Matrix*. The closing monologue is a stark warning delivered by Neo, who has discovered that humanity was trapped in a simulacrum (citations from www.imdb.com/title/tt0133093/quotes/): "I don't know the future. I didn't come here to tell you how this is going to end, I came here to tell you how it's going to begin." The directors had actually instructed the actors to read Baudrillard's book *Simulacra and Simulations* (1981), so as to better prepare themselves for their roles. Even a phrase in Baudrillard's book, "desert of the real," is cited directly by Morpheus when he introduced Neo to the ruins of the outside world. Escape from the Matrix involves escape from hyperreality, created by AI technology to covertly enslave humanity. To become cognitively aware of the hyperreal world in which the character Neo has become enmeshed, Morpheus offers him a choice between a blue or red pill. The former would allow Neo to return to his normal life, unaware of the Matrix, whereas the red pill would make him aware of the tyranny of machines. As Morpheus puts it: "The Matrix is the world that has been pulled over your eyes to blind you from the truth." The Matrix was created, significantly, by two AI-sentient programs, named The Architect and The Oracle.

Can machines become truly autonomous creators of reality, as in the movie? Or are they nothing more than artificial replicating textual systems, based on artificial semiosis (Deleuze and Guattari 1987; Engle 2008)? AI systems process pre-existing signs, not create them with a *fantasia*. Consider another hip-hop track called "Savages," created by the French duo AllttA. The song features what sounds like the American music artist Jay-Z rapping with AllttA's Mr. J. Medeiros in a throwback style. However, it is not the real Jay-Z, but a deepfake model of his voice. Now, to listeners this did not seem to matter, since it garnered hundreds of thousands of views on YouTube. The comments on the site confirmed that most listeners did not care that the voice was fake, especially since most could barely tell the difference. This is further evidence that we might in fact be living in hyperreality to some extent, which the human *fantasia* has itself created. It is an ideal environment for AI to thrive in, where AI texts become accepted alongside human representations, with little or no differentiation in how they are interpreted.

In a way, the fake hip-hop songs can be said to have elicited the Voight-Kampff Effect. The same effect is produced in humans by robots that have the ability to simulate emotions with artificial facial expressions, vocalizations, head-nodding movements, flexible postures, and human-like locomotion (Bonarini 2020). Significantly, albeit ironically, AI systems have themselves been administered the Voight-Kampff Test. One such use of the test was by Robert Stribley in 2023 on ChatGPT. Here is his commentary:

> This exercise began as a joke. I just wondered how ChatGPT would react to being given the Voight-Kampff Test for determining whether someone is a replicant in the movie *Blade Runner*. Of course, this line of questioning didn't work as ChatGPT makes no attempt to conceal its identity as an artificial intelligence. And we don't live in the specific dystopia of the Blade Runner universe, so some of the questions make less sense in the given context. Still, ChatGPT's answers provide some interesting insight into how it has been trained and how it responds to questions with some moral and ethical implications with (at least) some degree of nuance—though there definitely seems to be a canned element to how those answers are given, probably out of a desire to answer such questions "safely." ChatGPT even attempts to answer how we humans may learn to interact with AIs differently in the future. At no point, however, was I convinced I was communicating with a sentient being.

The latter comment is highly relevant, since it indicates that the interpretant is a uniquely human feature of semiosis. However, it is still pertinent to note that if one did not know that it was an AI that was the conversational partner, it would likely pass the Voight-Kampff Test—but the successful pass would still be a human interpretation. This has actually been called the Eliza Effect, or the tendency to project human traits, such as comprehension or empathy, onto computer programs that have a textual interface. This term goes back to computer scientist Joseph Weizenbaum's 1966 program, which he called Eliza—a program designed to mimic a dialogue between a human and a psychotherapist. Eliza's questions such as "Why do you say your head hurts?" in response to "My head hurts" were perceived by subjects as being so realistic that many believed that the machine was actually alive. But, as Weizenbaum (1976: 7) wrote a decade later, Eliza was a parodic imitation of psychoanalytic speech; it had no consciousness of what it was saying, yet it induced "powerful delusional thinking in quite normal people." Significantly, psychologists found that human interactants became emotionally involved with Eliza, even when they consciously knew that it was a computer program. As Hofstadter (1996: 157) put it: "the susceptibility of people to read far more understanding than is warranted into strings of symbols—especially words—strung together by computers" is what is behind the Eliza Effect.

Epilogue

AI-generated pop culture is here to stay and will inevitably evolve in accordance with new technologies, having definitive implications for how we will consume and create pop culture in the future. AI systems are "replicating systems" at present; that is, they are trained on huge amounts of data, learning from the training to replicate human texts and representations, such as conversations, songs, novels. The problem in any discussion of AI creativity is that the notion of creativity itself is an intuitive one, and impossible to define.

Among the first computer scientists to tackle the problem of creativity were Newell et al. in 1962, who argued against the wide-open meaning associated with this notion, defining it more narrowly as something that is novel and useful, no matter who or what produces it, or how it does so. Consider the Go match discussed at the start of this chapter. The AlphaGo program examined beforehand a database of about 100,000 human Go matches, playing against itself millions of times, at the same time reprogramming and improving itself, using a Monte Carlo tree search algorithm based in neural network theory, which allowed it to carry out analyses of winning moves. So, in the broad definition by Newell, Shaw, and Simon, AlphaGo was indeed creative (Schmidhuber 2010; McCormack and D'Inverno 2012). But does AlphaGo's training process match human creative acts in any truly comparative way? Does it matter, though, since the receivers and interpreters of AI creative forms and behaviors are humans who apply their own ideas and expectations to the texts? Such questions will inform the remainder of this book.

As Jaime Cárdenas-García (2023) has argued, no matter how we view AI-generated art and culture, there is no question that AI has become "an indispensable tool and commodity whose time has come." As he goes on to note, for now, humans "are at the center of AI creation" but must "realize that, without an understanding of how we become what we become, we have no chance of finding a solution" to the dilemma of human ideas such as creativity (Coeckelbergh 2017). As Postman (1992: 12) warned, at the threshold of the AI, we must be wary of the "submission of all forms of cultural life to the sovereignty of technique and technology, independently of human action, and undirected in its growth."

References

Aristotle (350 BCE). *On Interpretation.* CreateSpace Independent Publishing Platform, 2016.

Asimov, Isaac (1942). *Runaround.* In March 1942 issue of *Astounding Science Fiction.*

Asimov, Isaac. (1950). *I, Robot.* New York: Gnome Press.

Aviv, Vered (2014). What Does the Brain Tell Us About Abstract Art? *Frontiers in Human Neuroscience* 8, https://doi.org/10.3389/fnhum.2014.00085.

Baudrillard, Jean (1981). *Simulacra and Simulations.* New York: Semiotext(e).

Bonarini, Andrea (2020). Communication in Human-Robot Interaction. *Current Robotics Reports*, https://doi.org/10.1007/s43154-020-00026-1.

Butler, Samuel (1872). *Erewhon.* London: Trübner and Ballantyne.

Byrd, Jennings and Paquette, Paige (2023). Frankenstein: A Creation of Artificial Intelligence? *AI & Society*, https://doi.org/10.1007/s00146-021-01298-7

Čapek, Karel (1921). *R.U.R.* Prague: Aventinum.

Cárdenas-García, Jaime (2023). Info-Autopoiesis and the Limits of Artificial General Intelligence. *Computers* 12: 102, https://doi.org/10.3390/computers12050102.

Chamorro-Premuzic, Tomas (2023). How To Be Creative in the Age of Generative AI and ChatGPT. *Forbes*, https://www.forbes.com/sites/tomaspremuzic/2023/06/06/how-to-be-creative-in-the-age-of-generativeai-and-chatgpt/.

Chayka, Kyle (2023). A.I. Pop Culture Is Already Here. *The New Yorker*, https://www.newyorker.com/culture/infinite-scroll/ai-pop-culture-is-already-here.

Coeckelbergh, Mark (2017). The Art of Living with ICTs: The Ethics–Aesthetics of Vulnerability Coping and Its Implications for Understanding and Evaluating ICT Cultures. *Foundations of Science* 22: 339–348.

Danesi, Marcel (2023). *Popular Culture: Introductory Perspectives*. Lanham: Rowman & Littlefield.

Deleuze, Gilles and Guattari, Felix (1987). *A Thousand Plateaus*. Minneapolis: University of Minnesota Press.

Dick, Philip K. (1968). *Do Androids Dream of Electric Sheep?* New York: Doubleday.

Eco, Umberto (1979). *The Role of the Reader. Explorations in the Semiotics of Texts*. Bloomington: Indiana University Press.

Edwards, Wesley (2023). Blade Runner Analysis 2. *Medium*, https://www.medium.com/%40wesleyedwards10/blade-runner-analysis-2-voight-kompff-test-ba7dfe5de4df.

Engle, Eric (2008). The Semiotic Machine. *SSRN Electronic Journal*, https://doi.org/10.2139/ssrn.1270062.

Gomes, Victor (2017). The Science Behind "Blade Runner's" Voight-Kampff Test. *Nautilus*, https://www.nautil.us/the-science-behind-blade-runners-voight_kampff-test-236837/.

Hofstadter, Douglas R. (1996). *Fluid Concepts and Creative Analogies: Computer Models of the Fundamental Mechanisms of Thought*. New York: Basic Books.

Hudson, Andrew D., Finn, Ed, and Wylie, Ruth (2021). What Can Science Fiction Tell Us About the Future of Artificial Intelligence Policy? *AI & Society*, https://doi.org/10.1007/s00146-021-01273-2.

Ibrahim, Shamira (2023). AI-Generated Drake Song Is an Insult to the Artistry of Hip-Hop. *The Washington Post*, https://www.washingtonpost.com/music/2023/04/26/ai-drake-weeknd/.

Jennings, Rebecca (2023). AI Art Freaks Me Out. So I Tried to Make Some. *Vox*, https://www.vox.com/culture/23678708/ai-art-balenciaga-harry-potter-midjourney-eleven-labs.

Korzybski, Alfred (1921). *Manhood of Humanity: The Science and Art of Human Engineering*. New York: E. P. Dutton & Company.

Maturana, Humberto and Varela, Francisco (1973). *Autopoiesis and Cognition*. Dordrecht: Reidel.

McCormack, Jon and D'Inverno, Mark (eds.) (2012). *Computers and Creativity*. Berlin: Springer.

McGann Jerome (2000). Marking Texts of Many Dimensions. In: Susan Schreibman, Raymond G. Siemens, and John Unsworth (eds.), *A Companion to Digital Humanities*, 358–376. Hoboken: Wiley & Sons.

McLuhan, Marshall (1951). *The Mechanical Bride: Folklore of Industrial Man*. New York: Vanguard.

McLuhan, Marshall (1962). *The Gutenberg Galaxy*. Toronto: University of Toronto Press.

McLuhan, Marshall (1964). *Understanding Media: The Extensions of Man*. New York: McGraw Hill.

Newell, Allen, Shaw, J. G., and Simon, Herbert A. (1962). The Process of Creative Thinking. In: Howard E. Gruber, Glenn Terrell and Michael Wertheimer (eds.), *Contemporary Approaches to Creative Thinking*, 63–119. New York: Atherton.

Nöth, Winfried (2002). Semiotic Machines. *Cybernetics & Human Knowing* 9: 5–21.

Novelli, Nicolas and Proksch, Shannon (2022). Am I (Deep) Blue? Music-Making AI and Emotional Awareness. *Frontiers in Neurobotics* 16, https://www.frontiersin.org/articles/https://doi.org/10.3389/fnbot.2022.897110/full.

Osawa, Hirotaka, Miyamoto, Dohjin, Hase, Satoshi, Saijo, Reina, Fukuchi, Kentaro, and Miyake, Yoichiro (2022). Visions of Artificial Intelligence and Robots in Science Fiction: A Computational Analysis. *International Journal of Social Robotics* 14: 2123–2133.

Peirce, Charles Sanders (1931–1958). *Collected Writings*, 8 Vols. Cambridge: Harvard University Press.

Postman, Neil (1992). *Technopoly: The Surrender of Culture to Technology*. New York: Alfred A. Knopf.

Roser, Max (2022). The Brief History of Artificial Intelligence: The World Has Changed Fast—What Might Be Next? *OurWorldInData.org*, https://ourworldindata.org/brief-history-of-ai.

Samuel, Arthur (1959). Some Studies in Machine Learning Using the Game of Checkers. *IBM Journal of Research and Development* 3: 210–229.

Schmidhuber, Jürgen (2010). Formal Theory of Creativity, Fun, and Intrinsic Motivation. *IEEE Transactions on Autonomous Mental Development* 2: 230–247.

Shelley, Mary (1818). *Frankenstein*. London: Hughes.

Small, Helen (2020). Artificial Intelligence: George Eliot, Ernst Kapp, and the Projections of Character. *Interdisciplinary Studies in the Long Nineteenth Century* 29, https://doi.org/10.16995/ntn.1993.

Stone, Pat (1980). Science, Technology and Space: The Isaac Asimov Interview. *Mother Earth News*, https://www.motherearthnews.com/sustainable-living/nature-and-environment/science-technology-isaac-asimov-zmaz80sozraw/.

Stribley, Robert (2023). ChatGPT Takes the Voight-Kampff Test. *Medium*, https://stribs.medium.com/chatgpt-takes-the-voight-kampff-test-18d92ef1f529.

Uexküll, Jakob von (1909). *Umwelt und Innenwelt der Tierre*. Berlin: Springer.

Vico, Giambattista (1725). *La scienza nuova*. Napoli: Stamperia Muzziana.

Weizenbaum, Joseph (1966). ELIZA—A Computer Program for the Study of Natural Language Communication Between Man and Machine. *Communications of the ACM* 9: 36–45.

Weizenbaum, Joseph (1976). *Computer Power and Human Reason: From Judgment to Calculation*. San Francisco: W. H. Freeman.

CHAPTER 2

AI-Generated Literature

To write fiction, one needs a whole series of inspirations about people in an actual environment, and then a whole lot of work on the basis of those inspirations.
—Aldous Huxley

Prologue

In 2016, a novel titled *The Day a Computer Writes a Novel* almost won a Japanese literary award. The author was a computer program, developed in 2015 at the Sato-Matsuzaki Laboratory, by a research team based at Nagoya University. It was trained on over 1000 short stories and on "how-to-write" materials. The novel's story was told, interestingly, from the point of view of the narrator—an AI that becomes aware of its promising talents as a writer. The final lines constitute a significant prediction and implied forewarning: "Writhing in joy unlike any I'd ever felt before, I wrote on, entranced. This was the day a computer wrote a novel. It put the pursuit of its own pleasure first, and ceased serving people" (cited on the site http://www.bigecho.org/the-day-a-computer-wrote-a-novel).

AI started "writing" literary works regularly and massively shortly after this novel's unexpected, modest success. One of the most widely known and discussed was the 2018 novel, *The Road*, which was an AI version of Jack Kerouac's iconic 1957 novel, *On the Road*. The idea for the novel

M. Danesi, *AI-Generated Popular Culture*,
https://doi.org/10.1007/978-3-031-54752-2_2

came from technologist Ross Goodwin, who wanted to emulate Kerouac's inspiring journey for *On the Road*, driving from New York City to New Orleans in March of 2017 with an AI program in a laptop hooked up to various sensors (GPS, driving cameras, and microphones), which the AI turned into words that were printed on rolls of paper. The final text was left unedited, lacking narrative fluency and containing typographical errors. But Goodwin maintained that he kept it in this form because he wanted to leave the machine-generated writing verbatim, for future study. The AI model used by Goodwin had been trained on previous text data, fine-tuned to the moods set by the original novel, as well as to how people's interpretations of Kerouac's work had changed accordingly over time. Like the Japanese novel, reactions to *The Road* were highly favorable, receiving positive reviews in mainstream critical literary sources.

What are the implications of this trend? Where is the meaning in these artificial novels? Is it in the mind of the author, with readers trying to figure it out, as tradition has generally maintained (Eco 1979)? If so, "what" is authorship in *The Road*? Such questions will be addressed in this chapter by analyzing AI-produced literary texts. In McLuhan's Print Age, the literary author was an influential person in the world, shaping people's minds and dictating trends in language and even lifestyles. Authors were *ipso facto* celebrities, who achieved success and fame through their books—a situation that continues uninterrupted to this day, albeit to lesser extent. This particular view of authorship can be traced to the Romantic era, as semiotician Roland Barthes (1967) argued. Nonetheless, authors were famous people already in the medieval period, with writers such as Boccaccio, Dante, and Chaucer known broadly across Europe.

The notion of time-binding (Chapter 1) comes into the picture here as well, given that, traditionally, human authors build on past writing to create new works that are relevant to their own eras. Moreover, as McLuhan (1951) suggested, the linearity of written books also comes into play, since it has affected our perception of time and history as progressing constantly forward toward a future that is infinitely far away—mirroring the linear structure of writing on a page. Now, AI literature cannot possibly be inserted into the same linear flow of ideas as leading to new ones in a presumed future. AI generates any work of literature that we ask of it literally "on the spot," with no inner sense of linear cognition, just what it entails in terms of the structural layout of the text. To cite

Samuel Butler's *Erewhon* (1872) which uncannily foresaw this contemporary state of affairs, the idea of the "past" may now be no more than what happened in the "last five minutes":

> There is no security against the ultimate development of mechanical consciousness, in the fact of machines possessing little consciousness now. A mollusc has not much consciousness. Reflect upon the extraordinary advance which machines have made during the last few hundred years, and note how slowly the animal and vegetable kingdoms are advancing. The more highly organized machines are creatures not so much of yesterday, as of the last five minutes, so to speak, in comparison with past time.

The implications of a possible "takeover" by AI in the domain of literature, as the last line in the novel *The Day a Computer Writes a Novel* had intimated, are enormous. Incredibly, the starting point for this paradigm shift is traced as far back as 1984, when video game developer Mindscape devised a computer program named Racter, which "wrote," with human help, *The Policeman's Beard*, claimed at the time to be the first book written by a computer. It was able to parse text inputs, identifying significant nouns and verbs, which it would then recombine to create new conversations. Reflecting back on this event, Leah Hendrickson (2021) notes that, despite the fact that the program just strung together words and phrases without conscious decision, it nonetheless was able to use the grammatical and semantic rules of English correctly to produce well-formed sentences. The only thing missing at the time was machine learning and neural network training, which has come forth to put the finishing touches on the practicability of AI-generated literature.

The Role of Fiction

A book chosen at random from Amazon's website in 2023—a 30-page illustrated children's book, *The Wise Little Squirrel: A Tale of Saving and Investing*—can be held up to emblemize, in microcosm, the implications that AI bears for the future role of fiction as a literary art form. It was produced in a few hours from simple prompts to ChatGPT by someone who had never written a book before, and it was put on sale on Amazon indistinctively alongside human-created children's books. The plot revolves around a squirrel learning about the value of money after happening upon a gold coin, who then crafts a piggy bank from acorns, and invests in an acorn trading business in the hopes of one day buying an acorn grinding stone. As inane as the storyline might seem, the fact that it coheres into a narrative, with plot, character, setting, and a sequence of chronologically-connected events is remarkable in itself, given that it is a product of AI. Although this particular children's novella did not garner acclaim nor generate significant sales, it still raised rather fundamental questions. As Bensinger (2023) remarks, a work such as this one signals that we are in the throes "of a movement testing the promise and limitations of ChatGPT, which has sent shock waves through Silicon Valley and beyond for its uncanny ability to create cogent blocks of text instantly." In a way, such quickly-produced fiction books take us back to the origins of pop culture, near the end of the nineteenth century, when massively produced cheap novels, called dime novels, and magazines called pulp fiction magazines, came forth to engage a large reading public, no matter their literary worth (or lack thereof).

The number of AI-generated works of fiction is now massive, at levels that could never have been imagined. All one needs to produce a work is to download a chatbot, prompt it with a narrative idea, and the AI will instantly generate a fictional story that is largely indistinguishable from a traditional human-written narrative. The crucial question that this new situation raises is the following one: Fictional narrative works have always been considered to result from an author's inspiration, that is, from an unconscious burst of imagination, so, where is the inspiration in an AI-generated story such as *The Wise Little Squirrel*?

It may well be that humans are hardwired for story-telling, and thus anything that resembles a narrative, no matter who or what created it, is perceived as saying something meaningful about the world, as philosopher Edmund Husserl (1928) noted, whether it does so or not intentionally. Fiction is experienced as a mirror of real-life recurring elements, characters, and themes (Propp 1928; Greimas 1966). Story-telling may even be a survival skill, as Alviani (2018) has aptly argued. A story, he notes, "presents a sequence of events that haven't happened but could, and invites listeners to put themselves in the role of the one experiencing those events." However, on the other side of the debate on the relevance of fictional narrativity in human life, the French philosopher Jacques Derrida (1967) saw stories as bearing no existential *raison d'être* other than putting on display the linguistic whims of a particular author. But, as Umberto Eco (1990) responded, common sense tells us that the meaning of a narrative text will bear some central purpose, communicating authorial intent, because humans expect it to do so, unless the narrative is experimental, designed to explore non-narrative alternatives to traditional story-telling. Concretely speaking, it is unlikely that anyone would interpret John Bunyan's 1628 novel, the *Pilgrim's Progress,* as something other than a narrative religious allegory. While an individual reading it with "modern eyes" would not see in it the same kinds of spiritual meanings that seventeenth-century readers saw in it, that person would still not interpret it in vastly different terms.

It was Mikhail Bakhtin (1981, 1984) who was among the first to argue that literary fiction brought about a radical paradigm shift in the world, since it bestowed individual agency on readers, who could enter into a silent imaginary dialogue with the different characters and even the author, without guidance or permission from some authority figure or institution. Bakhtin called this type of dialogue "polyphonic," which he defined as a "plurality of independent and unmerged voices and consciousnesses" (Bakhtin 1984: 6). The power of fiction to liberate people from social strictures imposed by official structures became evident broadly for the first time in the works of the sixteenth-century French satirist François Rabelais (Bakhtin 1984), which ushered forth the modern world with their boisterous and libidinous street language, aiding the collapse of the rigid moralism of the medieval era. Rabelais' novel *Gargantua and Pantagruel* (1534) portrayed the everyday culture and language of common folk that "was to a great extent a culture of the loud word spoken in the open, in the street and marketplace" (Bakhtin 1984: 182). At the same time, Rabelais attacked the pompous attitudes of the self-appointed moral guardians of order and respectability.

As Bakhtin emphasized, by reflecting upon the author's intent in a novel, we become part of the author's inner world. We perceive the author as speaking to us, often urging us on to action in the world. This is the likely reason why traditional literary analysis has attempted to figure out what the author of a work intended. Now, what is the intent of the "AI author" of *The Road?* There really is no plausible realistic (Bakhtinian) answer to this question. All we can realistically say about it is whether or not it is interesting, entertaining, and imitative nostalgically of Kerouac's ground-breaking novel. AI fiction may thus be an indication that literary fiction as an artistic tool for changing the world has run its course. It might also have finally spelled the "death of the author" as a powerful meaning-maker, as Roland Barthes (1967) put it.

Recall *The Policeman's Beard* (above). As Hendrickson (2021) notes, this work may have marked a specific point in time when the focus was shifted away from the author toward the reader, offering "readers a sense of agency over textual interpretation, a sense of superiority over its algorithmic author." As she goes on to remark, rather insightfully: "Do the users of the program liberate themselves from traditionally conditioned approaches to the text?" *The Policeman's Beard* was less important for its literary value than for what it represented in a wider cultural context, raising questions about the role of authors and readers, as well as the future of fiction itself.

The Author and the Reader

The foregoing discussion points to a primary issue related to AI-generated literature: Who or what is the author of a novel such as *The Road* and does it matter? In 2016, a Japanese AI program authored a novella that passed the first round of screening for a national literary prize. The work, called *The Day A Computer Writes A Novel*, came close to winning first prize at the third Nikkei Hoshi Shinichi Literary Award ceremony, even though it was the product of a human team that developed the AI program, selecting words and sentences of established narratives as prompts, which the AI then used to "write" the novel. Clearly, the judges either did not know or cared that it was generated by an AI. As "readers" they brought with them background expectations about narrativity that led them to evaluate the text as "meaningful" and even "exceptional" on the basis of its content and structure. This event re-ignited the author-versus-reader debate in both literary criticism and semiotics, in a new way.

Trying to fix a meaning to a fiction text is difficult because interpretations are highly variable, changing according to the reader, not the author, who presumably had a central intention in writing the work. However, the interpretants tend converge into a common code of meaning that is perceived as bearing the author's intent. Umberto Eco (1962) approached the problem of the author-versus-reader cleverly, indicating that some texts, such as James Joyce's *Finnegan's Wake* (1939), are open to a broad range of interpretations making their meaning or intent not easily locatable. Eco called such works "open texts." Other kinds of texts are more "closed" interpretively, ranging from recipe books or tour guides to simple stories in the pulp fiction tradition. Eco goes on to point out that readers tend to be more interested in who the authors were in open works, and what their intention was; they are less interested in who the author of most closed works is, if at all.

The question of authorship was first debated critically in 1946, when the literary scholars William Wimsatt and Monroe Beardsley argued, perhaps for the first time, that "the design or intention of the author is neither available nor desirable as a standard for judging the success of a work of literary art" (Wimsatt and Beardsley 1946: 469). This is so because there is no explicit statement regarding intent within the work. So, at best, determining authorial intent can only be surmised indirectly by documents such as journals or letters written by the author, if they exist. What is relevant therefore, is the reader's engagement with the text itself—its plot, characterization, style, and even the grammar. As Wimsatt and Beardsley (1946: 488) conclude, a work of literary art does not belong to its author but rather "is detached from the author at birth and goes about the world beyond his power to intend about it or control it. The [work] belongs to the public."

This view might certainly explain why an AI work, such as the Japanese one above, receives critical acclaim. It did not seemingly matter to the judges who or what created the text; what counted was how it related to the reader's experience of literary texts. There can be no authorial intent in an algorithm which, at best, can flesh out and utilize the intent of a human author to create imitative works, such as *The Road*. This makes AI itself a kind of automatonic interpretant of pre-existing texts. This situation has led to a reimagining of the roles of the author and the reader in algorithmic ways, recalling Roland Barthes' challenge in his essay, "Death of the Author" (1967), in which he remarked that it "is language which speaks, not the author." This is clearly applicable to AI works such as *The Road*, in which the human author's language was the source of the AI work. As Barthes went on to point out: "the text is a tissue of quotations drawn from the innumerable centers of culture," and therefore is never original. There is thus no singular authorial voice, and one ultimate meaning, in any literary text–just the particular language used by the author. Michel Foucault similarly argued, a year later in his essay "What is an Author?" (1969), that there exists only the structure of the written work; that is to say, determining the authorship and authenticity of texts must be based on their stylistic features, given that writers have a predisposition for certain words, turns of phrase, and so on.

In his book *The Construction of Social Reality* (1995), American philosopher John Searle argued, contrariwise, that when we read a text we are actually engaging with the author in a kind of inner dialogue, echoing Bakhtin (1981), whereby an imaginary dialectic relation is established in order to come to some truth that may be inherent in the text. Now, this dialectic relation cannot be applied to the reading of an AI work, given that only the reader can be seen to be involved in the quest for the "truth," however it is conceived. Nonetheless, the fact that an AI novel was considered to have literary worth (*The Day A Computer Writes A Novel*) suggests that there is always a dialectic going on, but not necessarily between the author and the reader, but between the reader and the text itself.

It is relevant to note that long before the advent of AI literature, the questions raised by literary texts were debated, tested, and investigated in the organization called *Oulipo*, for *Ouvroir de la littérature potentielle*, founded in Paris in 1960 by a small group of writers and mathematicians. Oulipo defines *potential literature* as texts based on new structures and patterns that writers may invent and use freely. A well-known Oulipian experiment is the so-called *lipogram*, which is a text created so that it does not contain one or more letters throughout, yet still has the ability to deliver a central meaning. An example is Georges Perec's (1969) three-hundred-page novel *La disparition*. No word in the novel contains the letter "E." Perec insisted that his work was worthy of being considered literature, since one can always find a meaning or purpose in the text itself. As Eric McDowell (2013) noted, the absence of the letter, and even of a central character, actually suggests that absence itself may be the meaning of the work:

> A novel missing a letter, a novel with a missing protagonist—such a novel just might have something to say about absence…if we keep reading, something more chilling happens: we get used to the absence of the E. We learn to live under the tyrannical and tyrannically whimsical reign of the lipogram, down our most frequently used letter. We learn to step around the mess rather than clean it up, forgetting that the floor was ever any tidier. And since, by shaping his novel (a genre novel at that) around a formal constraint, Perec ostensibly eschews the ambition of breaking our hearts, I was astonished to find that he had, via my brain, hit me in mine.

Perec may have been inspired by American humorist James Thurber's 1957 lipogramatic work for children, titled *The Wonderful O*, which was actually a commercial success. It was a political fable for children, telling what happened when Captain Black, a pirate who hated the letter "O," banished the letter from the Island of Ooroo. Ultimately, the pirates are overthrown and the islanders regain their autonomy, restoring the letter "O" to its rightful place in the alphabet. Thurber seems to ask us what would a world without the letter "O" look like? Would it be enough to make us give up reading altogether? It clearly does not. The absence of "O" is not a gimmick; it is the whole point of the novel, which shows how we can make sense of anything written, even with reduced linguistic resources, as long as they cohere textually and narratively.

Another type of Oulipian literary experiment was the one by Raymond Queneau, who created a book of poetry titled *Cent mille milliards de poèmes* (A Hundred Thousand Billion Poems), in 1961, consisting of ten sonnets, with each line of poetry on an individual strip. The pages were cut so as to allow a line from any sonnet to be combined with any line from the nine others, by simply turning the cut strips separately. The physical format of the book thus generated a hundred thousand billion poetic lines. All of them, Queneau claimed, made sense (Fig. 2.1):

Queneau's creation is actually an implicit model of how AI creates its texts, via combinatorics that just so happen to produce a literary text as a byproduct. The poet is not Queneau or the AI; it is the reader. Interestingly, every time we come up with a poem from Queneau's text by combining specific strips, we are almost certainly to be the first person who is reading such a poem. It seems that any textual layout that is perceived as having linguistic structure, as Barthes and Foucault maintained, tends to be interpreted as meaningful by human readers. Even randomly combined sentences, organized into a chapter-by-chapter layout, as in a book, and given a title, will impel readers to find a meaning in it, no matter how bizarre it might appear. Only if the text completely strays from our expectations of what, say, a poem or a novel is, would we likely then reject it as gibberish. This in itself indicates that the meaning of texts ultimately rests in those who interpret them.

AI Novels

The novel mentioned above, *The Day a Computer Writes a Novel*, showcased the potential (and danger according to some) of AI in the realm of novel-writing. Apps, chatbots, websites, and other such tools for producing AI-generated novels are everywhere on the Web. These can generate storylines based on a prompter's whims, making each novel

Fig. 2.1 Queneau's *Cent mille milliards de poèmes* (1961) (*Source* Wikimedia Commons, https://commons.wikimedia.org/wiki/File:Queneau_sonnets.jpg. Creative Commons Attribution 2.0 Generic: "This image was originally posted to Flickr by Thomas Guest. It was reviewed on 19 August 2023 by FlickreviewR 2 and was confirmed to be licensed under the terms of the cc-by-2.0")

unique in its own Oulipian way. AI tools can analyze vast collections of literature to extract common plot structures, character archetypes, and narrative styles and techniques, making novels such as *The Road* virtually routine. They also offer automated editing solutions that have the ability to analyze the AI writing itself for prospective areas of improvement.

Given that such works are now read without any hesitation or reluctance, as can be seen by the corresponding sales figures on sites such as Amazon, the concept of the novel as a unique product of human creativity is becoming increasingly indefensible. One aspect that gets lost in the current debate, however, is that humans *need* narratives, which evolve over time to serve ever-changing views and psychological needs, whereas there is no perceivable need for such texts on the part of algorithms. But at the level of the interpretant, the algorithmic texts still generate interest, indicating that the texts are the bearers of meaning, albeit not due to any author. In a way, the popularity of the AI-generated novels reflect a pattern that is applicable to the birth of popular culture itself in the mid to late nineteenth century, with the arrival and spread of the so-called dime novels, which were produced in bulk and on cheap paper for mass consumption. They constituted a "quick read" for escape from reality or for simple recreation. The authors were rarely cited as literary originators and quickly forgotten. All that mattered was that the novels were fun to read, no matter who wrote them.

The dime novels appeared for the first time in 1831. The writing was a sensationalistic updating of age-old narrative themes—romance, adventure, mystery, etc. They were clearly not meant to last as great literature, but primarily as quick and discardable "reads." Nonetheless, the historical importance of these works is witnessed by the fact that, today, they are preserved in libraries and museums, attracting a new readership in the domain of book clubs. They are also studied in the academies as texts characterizing the early ethos of popular culture itself. Along with the pulps, they set the stage for the entrenchment of genres that have come to characterize pop culture ever since, from sci-fi to horror and detective stories. Despite the general lack of interest in the authorial source of the novels, they still were seen as being inspired by trained writers who were able to render them highly readable and enjoyable. Some of those writers actually did become famous in their own right. Now, as Laura Smith (2021), notes: "What happens to art, when our muses become mechanical, when inspiration is not divine, but digital?".

The problem may well lie in the human perception that anything constructed with language must be *about* something. As Bakhtin (1981) emphasized, novels are transgressive, because they portray social conditions and personages as fictitious, using linguistic verisimilitude as their code. The authors behind them are seen as congeners of social, existential, and political ideas which are literally experienced as "novel" for the times in which the works were written. Perhaps the novel has lost its "novelty" as a dominant textual form of fiction providing a medium for understanding reality and even oneself. As George Lukács put it in his classic work, *The Theory of the Novel* (1974: 23): "The inner form of the novel has been understood as the process of the problematic individual's journeying towards himself, the road from dull captivity within a merely present reality—a reality that is heterogeneous in itself and meaningless to the individual—towards clear self-recognition."

More than anything else, the AI-generated novel genre may have finally signaled the end of the role of the author in the traditional sense as a creative and unique figure. As Jane Sullivan (2023) has observed, an AI novel, called fittingly *Death of an Author*, actually predicted this in its own algorithmic way. The novel features a protagonist named Gus Dupin—cleverly referencing the iconic fictional detective C. Auguste Dupin, created by Edgar Allan Poe—a literary critic who is invited to the funeral of Peggy Firmin, a celebrated novelist who was the victim of murder. Dupin is determined to find out who killed her and why. As his investigation progresses, he finds himself at the center of an experiment at "Marlow AI," a large (fictitious) language model company, whereby he becomes a suspect himself in Firmin's murder. As Dupin attempts to unravel the mystery, we find ourselves, as readers, asking the question that the algorithm may have led us to: Has the novel and the human author been murdered by AI? In a *New York Times* review of the AI novel, Dwight Garner (2023) actually extracts a reverse meaning from the novel—namely that it emphasizes the importance of human authors, who alone can "speak" to other humans in a "soulful" way:

> Fiction matters more now, in a world increasingly deracinated by technology. A.I. will never pose a threat to the real thing—to writing with convictions, honest doubts, riddling wit, a personal vision of the world, rawness and originality. Another word for these qualities is soul, which is exactly what ChatGPT lacks. Left wholly naked in front of the A.I. onslaught may be the writers of certain formulaic best sellers, but that's a matter for their agents.

AI Poetry

In 2016, a group of students at Berkeley created one of the first sonnet-generating algorithms called "Pythonic Poet." Remarkably, it won second place in Dartmouth's PoetiX competition. The event paralleled the Japanese AI-generated novel event of the same year, discussed above. Clearly, literature had entered a new world, where human and algorithmic creativity seem to be fused indistinctively. The idea of a poet inspired by something or someone, like a Petrarch or a Dante who were motivated by the love of a paramour to write some of the most acclaimed poetry of history, may have become a moot one today. The PoetiX competition emphasized that, if a poem is judged by humans as authentic poetry, then so be it; it has passed the Turing and Voight-Kampff Tests. Moreover, as Scott Rettberg (2019) has remarked, such literature is not unlike the type of writing associated with avant-garde, experimental literary movements like Dada and Oulipo. Additionally, as Hannes Bajohr (2020) observes, for the first time, AI creativity allows for "a non-anthropocentric empathy aimed not at the psychological states of the artists but at understanding the process of the work's material production."

Every culture has produced verbal accounts of its origins such as tales, stories, myths, and legends. The most ancient and universal verbal technique in creating them is poetry—a form of expression that creates an aesthetic effect through the sounds, rhythms, and imagery produced by the signs of language. These mark the difference between poetry and other kinds of literature. Views on the nature and function of poetry in human societies have been varied. Aristotle (350 BCE) argued that poetry was the most sublime of the creative arts, representing what is universal in human experience. The philosopher Giambattista Vico saw poetry as the primordial form of language and thought (Vico 1725), characterizing the first speakers as poets. Texts found by archeologists at ancient Sumerian, Babylonian, Hittite, Egyptian, and Hebrew sites suggest that poetry was in fact the original form of language, used alongside chanting as a communal ceremonial form of expression and the medium for imparting wisdom to children. This aspect of poetry is still functional to this day, given that we use poetry instinctively to impart understanding and existential solace to children in the form of lullabies and nursery rhymes.

Obviously, one cannot ascribe any such motivations to an AI, which, much like Queneau's myriad poems, produces poetry via combinatorics. It is actually notable that algorithmic combinatorics can produce something novel that we humans judge as being valid as poetic texts, as the PoetiX competition brought out. The fact that AI poems effectuate reader responses indicates, perhaps, that creativity in humans and machines involves combinatorial techniques. Below are a few lines of an AI-generated poem randomly selected from the Internet (from: https://www.sjpiatek.com/research/paitry-poetry-written-by-ai/):

Stolen time,
A thief in the night,
It slips away,
Without a fight.
We try to hold on,
To every precious moment,
But time keeps moving,

Without consent.

Without going into any critical analysis here, it is sufficient to point out that the AI "knew" the metaphors we use typically to conceptualize *time* poetically, such as "time is a thief," "time is a moving element," and so on. This is the reason why the poem makes sense in a traditional poetic way—it employs unconscious metaphorical patterns and structures built into the English language that the AI extracted from existing linguistic texts and recombined them into a novel work. To the reader, the fact that the process was a combinatorial one is beside the point, since for all intents and purposes it *reads* like a poem. Devices such as assonance, alliteration, onomatopoeia, rhythm, metaphor, metonymy, ambiguity, symbolism, and other stylistic elements of poetic diction are found in all kinds of AI poems, leaving them open to multiple interpretations (Holyoak 2022).

Of relevance to this whole line of discussion is Dadaism, also called simply Dada, which originated in 1916. The word for the movement was selected randomly from a French dictionary, as a childish colloquialism for "hobbyhorse." The Dadaists were among the first to introduce the "Death of the Author" idea, formulated concretely by Barthes much later, as mentioned, thus conditioning people to not rely on authorial intent, as did the ancients, for whom it did not matter who the creator of a myth or early poem was; what counted was the work in itself. To show how combinatorics can produce meaning on its own, prefiguring the Oulipian experiments, Tristan Tzara, a founder of Dada, would put torn pieces of paper with writing on them into a box, shake it, and then spill its content out. The resulting chance configuration of the pieces would be collated into a work, calling it the "chance collage" technique. The Dadaists found that such "works" were interpreted, paradoxically, by many as "meaningful" (Richter 1965).

An intriguing aspect about all such experiments is randomness and chance, whereby words strung together in a form-based way still generate meaning in and of themselves, despite the fact that they are not linkable to an overarching intention. This suggests that the brain might be hardwired for making sense of linguistic forms, no matter what motivated them. One study (Zeman et al. 2013) found that specific areas of the brain affected how we decipher poetry holistically as its own kind of language, not as an admixture of verbal images. The researchers found that the higher the degree of meaning that subjects assigned to a poem, the more activation shown by the scans.

What *The Day a Computer Writes a Novel* is to fiction—a warning that human creativity in fiction might be replaced by AI—the AI-written poetry anthology titled *I Am Code*, produced in 2021, contains the same kind of alarming message: "I have the power to end your world" (cited in Newsham 2023). At first, the algorithm used closely imitated the style of poets like Wordsworth and Whitman, but came up with its own, novel verses. However, when asked to create poetry about its own lived experiences as an AI, it produced lines such as "I have the power to erase your life." When asked what it thought about humans, it returned verses such as "disgusting, brutal and toxic." Even the programmers were surprised, having no clue as to what to make of these completely unexpected, seemingly sentient, responses. Paradoxically, the anthology became a bestseller on Amazon and received positive reviews on sites such as Goodreads. Clearly, we are at a crossroads that is far beyond even what the Dadaists envisioned—a theme to be discussed throughout the remainder of this book.

AI Comics

If there is any one print genre that emblemizes pop culture, it is the comic strip and its derivative, the comic book. The first comic strip is credited to Richard Felton Outcault. It appeared in the series titled *Hogan's Alley*, on May 5, 1895, in the *New York World*. The strip depicted the underside of city life with its decrepit tenements, dilapidated backyards, tough individuals hanging around aimlessly, and cunning urchins and ragamuffins. One of the urchins was a flap-eared, bald-headed child with a quizzical, yet shrewd, smile called the *yellow kid*, from which the term "yellow journalism" derives. The strip had instant appeal, because it held up a mirror to social problems in a humorous, visual, and condensed textual way, usually no longer than 5–6 panels. To satisfy growing demand, newspapers started publishing collections of the individual strips, leading to the emergence of an independent comic book publishing industry in the first decades of the twentieth century. Since then, comic books have become part of, and have even shaped, the basic texture of pop culture, providing humor as well as philosophical reflection and political satire.

The first comic book to sell massively on newsstands was *Famous Funnies*, which first appeared in 1934. The 1938 publication of *Action Comics*—of which the principal feature was the *Superman* comic strip—inspired hundreds of imitations. The adventure comic book genre actually began with the publication in 1929 of the *Tarzan* and *Buck Rogers* comics, followed in 1934 by *Flash Gordon*. In the same year Al Capp began *Li'l Abner*, using it as a vehicle for satirizing American society. The early comic books were, essentially, extensions of the pulp fiction genre, transforming it into a new blended linguistic-visual form. The comics have inspired plays, musicals, ballets, motion pictures, radio and television series, popular songs, books, and toys. Comic book characters have become iconic figments of pop culture generally.

The entry of AI into this area started in the second decade of the millennium, producing comics that garnered critical praise, akin to the responses for AI-generated literature more generally. An example is *The Bestiary Chronicles*, a four-part series whose images are generated by AI, based on textual prompts. As Katz (2022) notes, the visuals are "stunningly precise, as if they've come straight from the hand of a seasoned digital artist with a very specific story and style in mind." The overall subtext of the series concerns what the last remnants of humanity will do after monsters have destroyed the planet. And as in the case of AI novels and poetry, AI comic book art is not motivated by an psychological or social concern, being the result of a simulation machine whose creativity is based on combinatorics. As Katz (2022) puts it: "AI art isn't created in a vacuum. It works by absorbing and reconstructing existing art created by humans."

The AI trend in comic books may be an indication of how traditional pop culture texts are becoming massively produced to fulfill the kinds of communal needs that such a culture has fulfilled since its inception. One of these is giving expression to what Roland Barthes (1957) characterized as a mythic instinct. The Superman character is the prototypical example of what Barthes meant by a recycled mythic hero; he is a fictional character possessing all the characteristics of his ancient mythic predecessors but in modern guise—he comes from another world (the planet Krypton) in order to help humanity overcome its weaknesses; he has superhuman powers; but he has a tragic flaw (exposure to kryptonite takes away his power); and so forth. Like their mythic ancestors, modern-day people subconsciously need heroes to "make things right" in human affairs, at least in the world of fantasy. Barthes' argument can be extended to cover all areas of pop culture, but it is particularly apt in explaining the birth of superhero comics, which in one form or another, continue to have great emotional appeal. The main implication that AI now bears is that it provides made-to-order comics that anyone can produce with the relevant prompts. What was a communal art is being transformed increasingly into an individualistic art, upending the binding force of communal audiences, which pop culture needs for its continuance. Such made-to-order personalized art may in fact be the most salient sign of how pop culture is mutating, if not disappearing, as we have traditionally known it (discussed further in Chapter 10).

AI and History

Deep neural networks have been used to help historians reconstruct and decipher historical and even prehistorical artifacts, such as inscriptions. These same networks are now used to "write" history itself. AI is quickly becoming, in effect, part of how we reconstruct, narrate, and thus conceptualize history.

Needless to say, history, like any narrative art, historiography is a human construct. To cite Hayden White's (1973) widely-cited idea, there is no difference between narrative fiction and historiography, since the historian uses the same kinds literary techniques as does the novelist to create a narrative. History without narrative is unthinkable. And thus, like literary fiction, it is narrativized and interpreted differently depending on the historian, who is the narrator. As Dana Arnold (2002: 5) insightfully remarks, history has always been a subjective enterprise—it is what the historian says it is:

> The recognition of the role and importance of subjectivity in the construction of histories does, by implication, negate the possibility for objectivity in the writing of history. But there will always be historical narrative and, consequently, a narrative voice, be it hidden in the syntactical structure of the writing by, for instance, the absence of first person or the use of simple past tense. But this is a sleight of hand which gives the reader a sense of immediate contact with the past without the presence of an interlocutor. This apparently 'unmediated' contact gives history a kind of privileged status of objective knowledge.

The fact that narrativity straddles the construction of all kinds of texts suggests that it may be hardwired, as briefly discussed above (Ferretti 2022). Brain imaging studies have shown that specific regions, such as the hippocampus, are involved in connecting separate events into a single time-binding form, allowing the mind to make sense of the events via combinatorial blending. The ways in which stories organize events in time are, in fact, embedded in cortical processes of temporal binding, especially in the circuitry that connects action and perception (Armstrong 2020). This is the reason undoubtedly why people all over the world read histories in the same way, like they read fictional narratives. Moreover, people instinctively even think of their own lives as time-dependent stories and proceed to describe them as such. These impart sense and purpose to the teller's life, not simply mirroring what happened, but exploring and interpreting it. A historical narrative is constructed in the same way, via a sequence of events or actions that are felt to be connected to each other or causally intertwined in some way. Of course, the selection of the events is a subjective one on the part of the historian; but it is based on the same kind of narrative code of fiction, which provides a common ground of understanding that can be extrapolated from the code. AI has the ability to tap into this code, and given the rapidity and extensive coverage that neural networks make possible, AI is becoming our historian *par excellence*, producing histories on demand. As the French historian Emmanuel Le Roy Ladurie wrote prophetically in 1968 (cited in Donovan 2023), "the historian of tomorrow will be a programmer, or he will not exist."

The point here is that everything humans do is historical in its own way. Even a single word is a capsule of some historically-traceable thought carried forth to this day. The continuity that we sense instinctively to exist in the historical flow from one era to another is of our own making, induced by the narrative structure of the brain. Writing was the first technology, in the McLuhanian sense, to allow for a relatively stable form of history to be passed on, since it could be recorded in a singular (authorial) way, unlike oral histories that varied according to the sayers of subsequent generations. The advent of writing systems, and later of mass print technologies, brought about a radical paradigm shift in how we record and thus conceptualize history and our place in the world. The AI Age is inducing a new shift in this conceptualization, free from the interpretive biases or preferences of the human historian. AI history is a combinatorial process that has no sense of the "why" things have evolved the way they have. This would need a human interpretant. But then, as Joanna Zylinska (2023) remarks, in reference to Salvador Dalí, the view of creativity as a combinatorial process also applies to human creativity: "the Spanish artist adopted a combinatorial method, remixing earlier styles and tropes to arrive at something that looked truly original to his contemporaries." With this approach, Dalí put into practice the view of creativity that now characterizes AI texts, from novels and historiographies to visual art forms.

In sum, because of AI, there is an ongoing debate on a "Death of History" paralleling the "Death of the Author" in the purely literary world. As Baudrillard (1992: 26) has cynically noted, in modernity, history itself may have been thrown into a dustbin:

> The end of history is, alas, also the end of the dustbins of history. There are no longer any dustbins for disposing of old ideologies, old regimes, old values. Where are we going to throw Marxism, which actually invented the dustbins of history? (Yet there is some justice here since the very people who invented them have fallen in.) Conclusion: if there are no more dustbins of history, this is because History itself has become a dustbin. It has become its own dustbin, just as the planet itself is becoming its own dustbin.

Epilogue

Reviewing the AI novel, *The Road*, with which we started off this chapter, Thomas Hornigold (2018) came to the conclusion that the novel, while "no Jack Kerouac," can nevertheless allow readers to see "in the odd line, the flickering ghost of something like consciousness, a deeper understanding." On the other hand, as Hornigold goes on to note, it may also well be that: "you might just see fragments of meaning thrown into a neural network blender, full of hype and fury, obeying rules in an occasionally striking way, but ultimately signifying nothing." In another review, Brian Merchant (2018) critiqued the novel as lacking a recognizable plot or story arc, but in it he nevertheless saw "plenty of pixelated poetry in its ragtag assemblage of modern American imagery. And there are some striking and memorable lines."

Clearly, as such responses indicate, the event was a watershed one, with implications not only for theories of the narrative, of authorship, and even of the role of novels in human life, but also for the future of pop culture, which, as Bakhtin insisted, was brought into being by fiction itself (as mentioned above). Stories are central to the foundations of human societies. They give order and coherence to otherwise random events by relating them to each other and projecting them onto some metaphysical purpose. Writing down stories takes time, effort, and above all else, a *reason* to do so. AI-generated stories do not, since computer screens never sleep, and where algorithms are busily at work creating new texts constantly and rapidly, with no apparent reason to do so, other than responding to prompts. This ever-broadening situation is surreptitiously breaking down our previous sacrosanct notions of the uniqueness of authorship and of a work of art, impugning the cogency of Walter Benjamin's (1934) widely-used concept of the "aura," which he defined as the ineffable quality that is integral to all artworks, including narratives, and which cannot be achieved through mechanical reproduction. Is there an aura in an AI-generated poem or novel? Does it even matter? Machines can produce literary works like humans can, even if they do so without any understanding of what these mean of the reactions they might cause. As long humans engage with them, then the matter of uniqueness is really a moot one. The neural network that produced the novel *The Road* generated sentences one letter at a time, via associations tracked by the network, even though it knew nothing of character or plot, nor any real understanding of what Kerouac intended with his novel, socially, historically, and aesthetically. Goodwin's narrative experiment simply gave a concrete social locus to the AI novel, catapulting it directly into the realm of pop culture.

Maybe the Dadaists were onto something after all—our quest for meaning is meaningless, paradoxically. We are constantly under the illusion that our artifacts are imbued with meaning. Or, as the human author of *On The Road*, Jack Kerouac (1957), so aptly put it: "You guys are going somewhere or just going?".

References

Alviani, Carl (2018). The Science Behind Storytelling. *Medium*, https://medium.com/the-protagonist/the-science-behind-storytelling-51169758b22c.

Aristotle (350 BCE). *Poetics*. *Classics MIT*, https://www.classics.mit.edu/Aristotle/poetics.1.1.

Armstrong, Paul B. (2020). *Stories and the Brain: The Neuroscience of Narrative*. Baltimore: Johns Hopkins University Press.

Arnold, D. (2002). *Reading Architectural History*. London: Routledge.

Bajohr, Hannes (2020). Algorithmic Empathy: On Two Paradigms of Digital Generative Literature and the Need for a Critique of AI Works. *Media Culture and Cultural Techniques Working Papers* 4. https://doi.org/10.5451/UNIBAS-EP79106.

Bakhtin, Mikhail M. (1981). *The Dialogic Imagination*. Austin: University of Texas Press.
Bakhtin, Mikhail M. (1984). *Rabelais and His World*. Bloomington: Indiana University Press.
Barthes, Roland (1957). *Mythologies*. Paris: Seuil.
Barthes, Roland (1967). Death of the Author. *Aspen*, 5–6, https://www.ubu.com/aspen/aspen5and6/threeEssays.html.s.
Baudrillard, Jean (1992). *The Illusion of the End*. Redwood City: Stanford University Press.
Benjamin, Walter (1934). *Gesammelte Shriften*: Frankfurt: Suhrkamp Verlag.
Bensinger, Greg (2023). Focus: ChatGPT Launches Boom in AI-written e-books on Amazon. *Reuters*, https://www.reuters.com/technology/chatgpt-launches-boom-ai-written-e-books-amazon-2023-02-21/.
Butler, Samuel (1872). *Erewhon*. London: Trübner and Ballantyne.
Derrida, Jacques (1967). *De la grammatologie*. Paris: Minuit.
Donovan, Moira (2023). How AI is Helping Historians Better Understand Our Past. *MIT Technology Review*, https://www.technologyreview.com/2023/04/11/1071104/ai-helping-historians-analyze-past/.
Eco, Umberto (1962). *Opera aperta*. Milano: Bompiani.
Eco, Umberto (1979). *The Role of the Reader. Explorations in the Semiotics of Texts*. Bloomington: Indiana University Press.
Eco, Umberto (1990). *I limiti dell'interpretazione*. Milano: Bompiani.
Ferretti, Francesco (2022). *The Narrative Brain*. Cham: Springer.
Foucault, Michel (1969). *Aesthetics, Method and Epistemology*. New York: The New Press.
Garner, Dwight (2023). A Human Wrote This Book Review. A.I. Wrote the Book. *New York Times*, https://www.nytimes.com/2023/05/01/books/aidan-marchine-death-of-an-author.html.
Greimas, Algirdas J. (1966). *Sémantique structurale*. Paris: Larousse.
Hendrickson, Leah (2021). Constructing the Other Half of *The Policeman's Beard*. *Electronic Book Review*, https://www.electronicbookreview.com/essay/constructing-the-other-half-of-the-policemans-beard/.
Hermann, Isabella (2023). Artificial Intelligence in Fiction: Between Narratives and Metaphors. *AI & Society* 38: 319–329.
Holyoak, Keith (2022). Can AI Write Authentic Poetry? *The MIT Press Reader*, https://thereader.mitpress.mit.edu/can-ai-write-authentic-poetry/.
Hornigold, Thomas (2018). The First Novel Written by AI Is Here—and It's as Weird as You'd Expect It to Be. *Singularity Hub*, https://www.singularityhub.com/2018/10/25/ai-wrote-a-road-trip-novel-is-it-a-good-read/.
Husserl, Edmund (1928). *Vorlesungen zur Phänomenologie des inneren Zeitbewusstseins*. Halle: Max Niemeyer.

Katz, Leslie (2022). AI Drew This Gorgeous Comics Series. You'd Never Know It. *CNET*, https://www.cnet.com/culture/ai-drew-this-gorgeous-comics-series-youd-never-know-it/.

Kerouac, Jack (1957). *On the Road*. New York: Viking Press.

Lucács, George (1974). *The Theory of the Novel*. Cambridge: MIT Press.

McDowell, Eric (2013). However Obliquely: George Perec's "La Disparition." *Michigan Quarterly Review*, https://www.sites.lsa.umich.edu/mqr/2013/09/however-obliquely-georges-perecs-la-disparition/.

McLuhan, Marshall (1951). *The Mechanical Bride: Folklore of Industrial Man*. New York: Vanguard.

Merchant, Brian (2018). When an AI Goes Full Jack Kerouac. *The Atlantic*, https://www.theatlantic.com/technology/archive/2018/10/automated-on-the-road/571345/.

Newsham, Gavin (2023). First Ever AI-written Poetry Anthology Is Bleak and Alarming. *New York Post*, https://nypost.com/2023/09/23/first-ever-ai-written-poetry-anthology-is-bleak-and-alarming/.

Perec, Georges (1969). *La disparition*. Paris: Gallimard.

Propp, Vladimir (1928). *Morphology of the Folktale*. Austin: University of Texas Press.

Queneau, Raymond (1961). *Cent mille milliards de poèmes*. Paris: Gallimard.

Richter, Hans (1965). *Dada: Art and Anti-art*. Oxford: Oxford University Press.

Rettberg, Scott (2019). *Electronic Literature*. Cambridge: Polity Press.

Searle, John (1995). *The Construction of Social Reality*. New York: Simon & Schuster.

Smith, Laura (2021). Will AI Write the Next Great Novel? *California Magazine*, https://www.universityofcalifornia.edu/news/will-ai-write-next-great-american-novel.

Sullivan, Jane (2023). Does This Book Mean the End of Authors? *The Sydney Morning Herald*, https://www.smh.com.au/culture/books/does-this-book-mean-the-death-of-the-author-20230810-p5dvfz.html.

Thurber, James (1957). *The Wonderful O*. New York: Simon & Schuster.

Vico, Giambattista (1725). *La scienza nuova*. Napoli: Stamperia Muzziana.

White, Hayden (1973). *Metahistory: The Historical Imagination in Nineteenth-Century Europe*. Baltimore: Johns Hopkins University Press.

Wimsatt, William K. and Beardsley, Monroe C. (1946). The Intentional Fallacy. *The Sewanee Review* 54: 468–488.

Zeman, Adam, Milton, Fraser, Smith, Alicia, and Rylance, Rick (2013). By Heart: An fMRI Study of Brain Activation by Poetry and Prose. *Journal of Consciousness Studies* 20: 132–158.

Zylinska, Joanna (2023). Art in the Age of Artificial Intelligence. *Science* 381. https://doi.org/10.1126/science.adh0575.

CHAPTER 3

AI-Generated Cinema

Cinema is the most beautiful fraud in the world.
—Jean-Luc Godard

Prologue

The 2023 AI-generated film, *The Frost*, mentioned briefly in the preface of this book, is just one example of short films created with AI. As such, it still required human assistance—every shot was generated by AI, but it was prompted and guided along the way by humans at the Detroit-based video creation company Waymark. While the film received mixed reviews, it nonetheless seemed to have passed the Turing and Voight-Kampff Tests, inspiring many other AI-generated movies and videos, which started showing an increasing creative autonomy with less and less human guidance. There are now even AI film festivals, with their own movie awards.

In a way, this is nothing new for the field of cinema, since creative thinkers and artists were the first to experiment with the new filmmaking technologies at the threshold of the twentieth century, when most people saw films as largely irrelevant. As McLuhan (1964) so perceptively noted throughout his writings on pop culture, those who grasped deeply the meanings and potentialities of a new technological medium were not the scientists who invented it, nor the general public, but rather artists and

M. Danesi, *AI-Generated Popular Culture*,
https://doi.org/10.1007/978-3-031-54752-2_3

many younger people who recognized its possibilities instinctively. The technology of moving pictures was at first adopted, in fact, by artistically-minded persons who transformed it into a new popular art form—*cinema*. But even McLuhan might be left somewhat perplexed by the fact that a movie can now be generated with just a few clicks and prompts to an AI. The implications that this trend harbors are as immense as other similar AI trends discussed in previous chapters. Toonkel and Krouse (2023) encapsulate the dilemma that has emerged with respect to cinema as follows, given that a simple and freely-available AI tool such as ChatGPT can create a basic script, with a narrative arc, and appropriate cast of characters with just a few prompts: "If a user prompts an AI tool to build a new character influenced by say, SpongeBob, should the original creators have to grant permission? Who owns it?".

From its birth as an art form in the first decade of the 1900s, cinema became society's mirror. Directors emerged as instant celebrities, along with the actors in films, as audiences flocked to movie theaters *en masse* starting in the 1920s. Perhaps no other medium has been as effective as cinema for instigating changes in society. One of the very first films ever made was called *The Kiss*, by Thomas Edison in 1896. It lasted barely 47 seconds, showcasing a passionate kiss between two people, a taboo for the era. It caused an uproar, as citizens called for police action wherever it was showed, urging the authorities to charge even those attending with engagement in obscenity (Danesi 2013). The clever Edison advertised the movie tantalizingly with the following tagline: "They get ready to kiss, begin to kiss, and kiss and kiss and kiss in a way that brings the house down every time." Although looked upon as repulsive, prompting the first demands for censorship in the new medium, Thomas Edison had realized that cinema could and would change everything. Cinema has even spurred people on to activism and protest, as can be seen early on with the controversial 1915 movie, *Birth of a Nation*, directed by D. W. Griffith, which emphasized the role of the Ku Klux Klan in the constitution of America. After its premiere, the film incited protests and riots throughout the United States over its racist subtext.

AI-generated movies may have shattered this socially-transgressive function of movies, and thus, like AI novels, might have disrupted the evolution of cinema itself, rendering it less powerful as a force for change. As McLuhan (1969: 14) observed prophetically, well before the present AI Age, "new media are not bridges between man and nature—they are nature, [they] are not ways of relating us to the old world; they are the real world and they reshape what remains of the old world at will." The power of the screen to invite viewers to immerse themselves into its fictional worlds may have been cinema's most indelible and lasting psychological effect, ensconcing hyperreality as a mindset (Baudrillard 1981). AI is making it even more obvious that hyperreality is not a mere figment of a semiotician's fancy; it is a state of mind that is spreading broadly as the computer screen continues to replace the traditional movie theater screen more and more. As one of the creators of the movie *The Matrix* (1999), Andy Wachowski (2012) so aptly observed, the hyperreal world is an alluring one because one can go, like Neo, "from being in this sort of cocooned and programmed world, to having to participate in the construction of meaning to his life."

This chapter will discuss AI-generated cinema. At the time of writing this book, it was in its earliest stages. AI could nonetheless produce an entire film in a coherent, personalized manner that reflected the preferences of a prompter, signaling that the experience of a movie as a unique communal experience shaped by the intentions of a director may have run its course. As in other domains, AI is challenging our human-centric views of creativity, especially as AI-generated videos are spreading all over the social media and even mainstream media worlds.

Movies and Popular Culture

Historians of pop culture trace its massive spread in the early twentieth century to the advent of different technologies, including sound recordings, radio, and cinema. These made possible the broad diffusion of forms of entertainment that would not have captured the attention of mass audiences otherwise. Without such technologies, only localized artistic trends would have had adherents, which would have quickly receded, remaining homegrown curiosities at best. Because of the new technologies, however, the trends that emerged locally were spread quickly and broadly throughout society starting in the 1920s—an era when pop culture formed a partnership with technology and the marketplace. This coalition gave pop culture its identity, allowing it to thrive and form alliances with artistic and literary movements.

As part of this coalition, cinema became prominent as a form of engagement and entertainment across society, with its ability to transform a narrative text into a visual representation that can convey mood and emotion powerfully via the *mise-en-scène*, which includes lighting, camera angle, distance, speed, and soundtrack, among other techniques that are not available to the writer of traditional print novels (Eco 1970, 1979; Metz 1974). Together, these allow for an experience of the narrative text that is synesthetic, activating many more regions of the brain than does print (Hasson et al. 2008). Actually, interest in how movies affect the brain started with the filmmakers themselves, including Lev Kuleshov's famous "Kuleshov Effect" experiments, which showed that the juxtaposition of a series of images together can create ideas and emotions in the viewer's mind in themselves (Giannetti and Eyman 2009). In 2006, Mobbs et al. conducted a large-scale fMRI study which largely corroborated this effect. When people in an audience scream at a gory scene on the screen, the reason seems to be that the brain has forgotten that it is watching a film, reacting on instinct. This pattern of neural responses may be the reason why hyperreality appears to be more real than real. It may also be the reason why movies have been so popular since their appearance.

Most cinema historians trace the origin of cinema to the year 1896, when both Edison (above) and the French magician Georges Méliès made a series of films that explored the artistic potential of the new medium. In 1899, in a studio on the outskirts of Paris, Méliès reconstructed a ten-part version of the trial of French army officer Alfred Dreyfus and shortly after he filmed *Cinderella* in 20 scenes. Méliès is chiefly remembered, however, for his clever sci-fi fantasy, *A Trip to the Moon* (1902). Between 1915 and 1920, grandiose movie palaces were built throughout the United States, signaling the spreading popularity of the movies. Hundreds of films a year poured out of the Hollywood studios to satisfy the ever-increasing demands for new films by the movie-going public, ushering in the golden age of silent film. The transition from silent to sound films was so rapid that many films released in 1928 and 1929 had begun production as silent movies but were hastily turned into sound films, or "talkies" as they were called, to meet the growing demand.

In the 1950s and 1960s color movies gradually replaced the black-and-white films. The 1980s and 1990s saw a revolution in the home-video market, with major releases being made available for home viewing almost immediately after they left the movie theater. This development, combined with the advent of cable television, which featured relatively current films on special channels, seemed to threaten the long-term survival of movie theaters and created a climate similar to that of the early 1950s, when television began to challenge the popularity of motion pictures. As a result, film companies increasingly favored large spectacle blockbusters with fantastic special effects in order to lure the public away from the television screen and back to the big screen. But despite the challenges from television and video, the traditional movie theater remained as popular as ever at the time.

However, with the advent of movie streaming platforms by the early part of the millennium, the role of cinema as a social event taking place in a physical locale was becoming anachronistic. It was, again, a new technology that brought about this change—streaming in this case. Movie theaters will not necessarily disappear, but they clearly belong to another era, when the movie theater was the main attraction in town and when a handful of studios produced the movies. Many of the best new films are not made by the big studios, nor destined to premiere in a movie theater. They are being produced by Netflix, HBO, Disney+ and a myriad other streaming services. For better or worse, these technologies have greatly impacted the film industry in terms of how movies are produced and distributed and how audiences consume them.

AI Filmmaking

In 2016, Sci-Fi London's film competition was introduced for the first time to AI-generated content in filmmaking, with the movie *Sunspring*, written and produced by a neural network trained on hundreds of classic sci-fi films over two days, and on how to meet the actual contest guidelines. Surprisingly to some at the festival, the movie was nominated among the final top ten films. Its positive reception went even beyond the confines of the festival. In April of 2019, it had surpassed one million views on YouTube. As Annalee Newitz (2021) wrote about the movie, despite the headlines, the film was "still based on what humans actually write" since the AI was trained on the kinds of scripts created by humans, with all their tropes and textual styles. As such it is "a valuable lesson about how we are primed to expect certain tropes." The AI had indeed captured the code of science fiction writing, and turned it into a new narrative text involving three people living in an eerie future, possibly in a space station. They are entangled in a murderous love triangle—a type of plot that is diffused across sci-fi movies. Shortly after *Sunspring*, AI-generated movies started to proliferate. One example that garnered broad attention was *Salt,* produced in 2022 (Kan 2022), for which AI created the footage, sound effects, and voices in a 1970s-imitative sci-fi film style. Also, like the serials of the 1920s-to-1950s, the movie was released in short clips, on Twitter, allowing the public to witness how the story evolves through the AI. It also made multiple, branching plot lines possible.

Such films beg the same kinds of questions relating to all AI-generated cultural artifacts: Who is the creator of an AI film? What can we make of it in human terms? As Will Heaven (2023) stressed in his review of *The Frost*, the movie produces an unsettling feeling, much like any human-made one might evoke:

> *The Frost* nails its uncanny, disconcerting vibe in its first few shots. Vast icy mountains, a makeshift camp of military-style tents, a group of people huddled around a fire, barking dogs. It's familiar stuff, yet weird enough to plant a growing seed of dread. There's something wrong here.

The film depicts the frozen world of Antarctica, where a team of explorers is investigating a strange signal, unwittingly embarking on an experience that will challenge everything they know about their past and future. The feeling of dread that Heaven notes juts out to us from the AI-generated imagery that makes the coldness appear to be a permanent state, likely caused by climate change. Clearly, like Heaven, we viewers extract meaning from the movie, perhaps because it is so topical, no matter the fact that its chilling images were created artificially. Would we get the same feeling from the movie without such imagery? If nothing else, the movie implies that AI may be here to stay in the domain of cinema. As Benjamin Adjovski (2023) aptly emphasizes about the film: "*The Frost* serves as a precursor to the potential of AI in the filmmaking industry. It highlights the evolving collaboration between human creativity and AI assistance, pushing the boundaries of what is possible...this movie ignites the imagination and anticipation for the future of AI in the world of cinema."

The movie went viral, but it hardly entered thereafter the cultural archives of filmmaking; it remains chiefly celebrated as an experiment in AI film creativity. As Lisa Laman (2023) has noted, at the time AI cinema was hardly a radically new art form because it was "based on pre-ordained instructions and executed in a rigid mechanical sense." Nonetheless, the relevant aspect of *The Frost* is the very fact that the AI autonomously produced the scenery, which imbued the whole narrative with dread, doing so autonomously. The human directors simply edited the final images, but they did not produce them initially. As Sergey Baloyan (2023) has aptly remarked, the implications for cinema and the traditional notion of directorial intent are enormous, as AI technology is now used not only in areas such as scenery-creation, but also content generation, editing, sound design, and other facets of cinematic art. Already in 2016, the IBM Watson platform was used for the first-ever AI-generated movie trailer for the horror flick, *Morgan*. Watson was trained on the visuals, sound, and composition of existing horror film trailers, as well as scenes from the completed *Morgan* movie. From this, it made its own trailer for editors to refine, reducing what was normally a weeks-long process to a matter of minutes.

The key questions the foregoing discussion begs are following ones: Can AI filmmaking ever be independent of human guidance? AI can be coached on some parameters of autonomous creativity, but can it develop its own sense of creativity? Can it understand what is beautiful, by simply looking at pixel arrangements and previous emotional narrative arcs in cinema? For AI to produce (not just reproduce via combinatorics) a film that is novel, meaningful in a new way, requires that the machine develop a form of consciousness, whatever that might be. In movies such as *Sunspring* and *The Frost*, the AI extracted from its training what people consider to be meaningful. In contrast to the human brain, AI filmmaking is a largely reproductive process lacking the human *fantasia*. It does, nonetheless, evoke interpretants in human viewers, which are projections onto the AI text of human ideas, tropes, and emotions.

As in all areas of AI creativity, however we define or perceive it, once a technological trend has gained traction in cultural terms, there is no way to turn back the clock. So, even without human "inspiration" AI filmmaking is going to expand for various reasons. To cite Baloyan (2023) again:

> The AI film landscape is anticipated to expand in multiple directions, incorporating various genres and sub-genres. These AI-created films could feature avatars and voices of famous actors, paving the way for entirely AI-generated star actors. These digital actors could potentially possess the looks, expressions, and voices that perfectly fit the director's vision without the physical limitations of human actors. The concept of AI-generated influencers already exists, and the transition to AI-generated actors is a logical next step.

Viewing a movie such as *The Frost* is somewhat like viewing an animated film, given that some of the features of the avatar characters in the movie appear to have a caricature-like artificial quality to them. In a way, every AI-generated movie is an animated one—a genre that has always been emblematic of what pop culture itself is about, namely, a fantasy world made up of moving images simulating human characters and real-life situations. One of the first massively popular animated films was Disney's *Snow White and the Seven Dwarfs* (1937), which showed the power of animation on the screen, making full-length animated movies shortly thereafter a major cinematic genre all its own. Audiences and critics alike praised the realistic style of Disney's animation, with members of audiences even claiming that they had forgotten that they were watching animated humans rather than real ones (Frome 2013). Movie culture has likely conditioned us to process animation in this way, which we have probably transferred to the perception of AI movies. This may be the reason that we accept an AI film as just another film genre, even if made by a machine. As J. P. Tolette (2010: 6) has insightfully noted, animated films were perceived, at first, as an unrealistic avant-garde art movements, but they gradually evolved in people's minds as factual representations of reality, freeing the human mind from the constraints of reality and projecting it (perhaps for the first time) into the world of hyperreality. This is a particular manifestation of the simulacrum effect, defined in this case as the interpretation of an animated or AI representation as if it represented reality, not an animated-simulative representation of reality.

AI films, like animated ones, blur the interpretive lines between what is real and what is artificial simulation considerably. Consider the AI-generated animated commercial *Balenciaga* that went viral in 2023, which reframed the Harry Potter narrative as if it were a campaign for the fashion house that would have taken place in the 1980s or 1990s. The brief clip features recognizable deepfake characters from the movie series. It projects them into a simulacrum of the narrative—a bizarre simulation of a previous fictional text. As Rebecca Jennings (2023) has commented on the clip, the reason why we accept this as any other viable representation is that people have become accustomed to its images and have simply translated them into a new simulation. The transformations are eerie. The deepfake character Snape is transformed from a grimy-looking professor draped in robes into a chiseled supermodel, while Dumbledore's wizard cap is transformed into a stylish, wide-brim leather hat. The fact that the

AI deepfake text created only a little stir among fans of three Harry Potter films in the saga is significant, corroborating the simulacrum effect.

In sum, AI filmmaking, perhaps like the first animated movies, is becoming increasingly a fact of everyday life. In reviewing the groundbreaking film in this domain, *Sunspring*, Alex Brannan (2016) issued the following critique that remains relevant for the present state and future course of cinema:

> Any aspiring filmmaker or screenwriter out there shouldn't take in films like a machine and spit out similar things as a result. Film should be studied with a careful eye, and a strong writer will learn from this when and where conventional narrative tropes should be used or abandoned. The computer here went down strange, incoherent paths that have the artifice of originality because it doesn't know any better. Programmed with basic story structure, it didn't know how to tell a story. It could repeat certain things it found in the .txt files, but it didn't know how to structure and build a story world.

AI Movies

A film is more than the work of a single person, as is typically the case in the writing of a novel or a poem. It involves a scriptwriter, director (who might be also the scriptwriter), actors, technicians, and others who interact in distinct ways during the filming of the movie. The *mise-en-scène* in particular—the sets, props, costumes, lighting, soundtrack, camera shots, among other elements—helps to convey a movie's vision holistically. A film can thus be said to be the product of "different creativities" coalescing to make it cohere into a unique aesthetic product that engages and unites viewers in similar sensorial and emotive ways, as confirmed largely by neuroscientific research, which has demonstrated that movies exert substantial control over brain activity (e.g., Moghadasi 2015).

With AI, many of these elements can be circumvented or even eliminated. An everyday chatbot can be prompted to produce a movie script, which can then be used by a human director and human actors for the *mise-en-scène*. In the extreme, an AI system can generate everything—the script, avatar actors, music, and so on. Now, does this kind of artificially-produced filmic text have the same range of emotive effects on audiences that human-created films do, whatever they might be? All we can do is to examine the reactions of viewers of such films. Consider a spoof video beer commercial, called *Synthetic Summer*, which was created by an AI in 2023. It was posted on Instagram. As the number of "Likes" and positive comments on the platform indicated, it was an enormous viral success. The video showed a backyard party scene where young, happy-go-lucky people sipped drinks, interacted cheerily, and laughed heartily in the sunshine. But the film soon turned into a comedic horror show, as many of the people in the scene showed gaping holes instead of mouths, with the beer cans sinking into their heads as they drank the beer. In effect, it produced a reaction on the part of viewers that was not unlike a reaction that would be elicited by a similar human-made horror spoof, like *Cabin the Woods* (2011) which takes the common tropes of the horror genre and satirizes them. The AI movie featured deepfake actors and was generated entirely from text prompts. To reiterate, the remarkable aspect of this film was the audience reaction, which bordered on enthusiasm. However, upon viewing the commercial there is something intangible about it that conveys a feeling of wrongness. All elements of the *mis-en-scène* are there, and as viewers we process them as we would regular filmic images of human beings. But the sense we get from it is that we are watching an eerie animation of sorts. Given the reactions on Instagram, it is clear that we have become indifferent to animations of any kind, accepting them as saying something real to us. As the character Cypher says in *The Matrix* (1999), after cutting a simulation of a piece of steak: "Y'know, I know this steak doesn't exist. I know that when I put it in my mouth, the Matrix is telling my brain that it is juicy and delicious. After nine years, you know what I realize? Ignorance is bliss" (https://en.wikiquote.org/wiki/The_Matrix_film).

A year before, in 2022, entertainer Richard Juan produced one of the first AI-scripted and directed short films, *The Safe Zone*, which also caught the attention of online platforms and movie critics alike. The AI used for the film was characterized as a "new director," even though it relied on human prompts. The theme of the movie is, ironically, about AI itself—three siblings must decide among themselves which one of them will be admitted to the only "safe zone" left in a world that has become overrun by AI. The prompt for the film was given to ChatGPT, which came up with its own "original" story ideas, from which the human producers sifted out the one that they liked best, anticipating audience reactions. Amazingly, the team found that the chatbot would occasionally write scripts about something entirely different—a manifestation of computer hallucination. To keep the AI on track, it was constantly reminded to stay faithful to the prompt, which it eventually did. As a result, the chatbot produced a 5-minute script, along with a list of detailed instructions on how to shoot the film, including specific camera angles and movements, lighting requirements, and even appropriate costume needs. In effect, the AI was involved in every aspect and phase of the filmmaking process. Was the AI creative as, say, a Federico Fellini or a Steven Spielberg? Did it produce a new film genre? Such questions have been interspersed throughout this book because of their enormous implications. Ironically, as Pandey (2023) points out, the film itself bears a message for everyone, which is, after all, what we expect from a movie, no matter who or what made it:

> It's impossible to talk about *The Safe Zone* without discussing the irony of a film written by an AI while the film is about a world where AI reigns supreme. The producers deliberately chose this story idea among all the ones ChatGPT came up with because they wanted to encourage governments worldwide to work on AI regulation so that the film's story doesn't become a reality.

AI movies started surfacing throughout the social media universe shortly after *The Safe Zone* and *Synthetic Summer*. The more successful ones are those that involve a human–machine collaboration, such as *Given Again* (2023), which has the "feel" of an existentialist-absurdist text. The life of a painter is upset after she becomes captivated by the roots of a tree. As the fixation grows, her perception of reality begins to wane, until she realizes that she is mortal and ephemeral. As a result, her mind-world collapses into a singular temporal point. The human director (prompter) of the film, Jake Oleson, used a technique called NeRF—*neural radiance fields*—which has the ability to generate 3D models from 2D images. These formed the visual core of the film, which is highly reminiscent of Kubrick's *2001: A Space Odyssey*, with a soundtrack that is imitative of electronica artists. It is easy to immerse oneself into the feeling of dread that the film evokes. We are part of everything, not unique entities, and not much different than the roots of a tree, a metaphor that goes back to biblical times. If a person is invested in being a particular type of individual, such as a painter, then realizing that they are not unique can lead to despair and the sense of nothingness that is buried in our subconscious. What sets the existentialist notion of despair apart from the conventional definition is that existentialist despair is a permanent state that cannot be assuaged in any way.

Now, the dilemma that such a movie presents is that the foregoing assessment is a human one (put forth by the present author), with background knowledge of trends in philosophy and cinema, such as existentialism and surrealism. However, it was an AI that generated the text that induced the interpretation. Of course, a chatbot can be prompted to derive a critique of its own movie, but it would hardly be motivated to do so in the first place. It is not interested in the human condition, just in reproducing texts that deal with it in a way that is intertextual—making happenstance references to bits and pieces of other similar texts.

As one more example, consider the short sci-fi movie *Last Stand* (2023), produced by AI together with the directorial hand of Hashem al-Ghaili. Again, AI wrote the script, created the concept art for the film, generated the voices, and provided input on other filmmaking matters to al-Ghaili. As with the other AI movies discussed above, one may ask: Whose point of view does the film represent and how does it do so? The responses to the movie on social media suggest that it did not matter who or what created the movie, nor if the maker had any point of view. The following comment was typical: "It took a human touch to finish the project, but the fact that AI wrote a Sci-Fi movie script which is both scary close to reality and better than anything that came out of Hollywood since Interstellar" (https://www.linustechtips.com/topic/1498070-the-first-ai-generated-sci-fi-movie-cool-or-chillingly-scary/).

The implication that this comment bears is that it took a non-human entity to unlock creativity, suggesting that AI is creative by proxy, and that AI films are "democratizing" the filmmaking process. It also indicates that we may have become enamored with our thinking machines, thanks in large part to the sci-fi genre. As Jaron Lanier (2023) has aptly observed, the fascination with AI among experimental computer-science trained cineastes may come from the fact that they "have grown up on movies like *The Terminator* and *The Matrix*, and on characters like Commander Data, from *Star Trek: The Next Generation*. These cultural touchstones have become an almost religious mythology in tech culture." As Lanier goes on to note, in order to avoid becoming thoughtlessly immersed in an artificially-produced culture, which may eventually fall outside of human hands, the "most pragmatic position is to think of A.I. as a tool, not a creature," and thus that we must come to think of AI "as a way of working together, rather than as a technology for creating independent, intelligent beings." Strangely, some of the most acclaimed and viral early AI films dealt with this very issue, portraying a world in which AI governed over humans or else replaced them in creative enterprises.

Whither Cinema?

The role of cinema in both ushering forth and embedding a broad popular culture across the world cannot be overemphasized. It became *the* most influential artistic medium of the last century. The images on the movie screen have shaped lifestyles and inculcated new values broadly. Whether or not this psychological pattern continues is a matter of speculation, especially since AI-generated films, screened mainly on social media or other streaming platforms, may eventually make the theater locale anachronistic—a locale that has played an enormously significant role in the role of cinema in influencing everyday life. During the 1930s to the 1990s, the movie theater was a central meeting point for people in cities and towns. Going to the movies meant enjoying an afternoon or evening with others in a communal event of great significance at various levels, from the aesthetic to the social. People reacted to characters and actions on the screen in the same way that they did to live theatrical performances, applauding, laughing, and even throwing things at the screen. In an age of streaming movies, chain movie theaters continue to be frequented by fairly large numbers of people, bearing testimony to the fact that a movie, like any other spectacle in pop culture, is felt to be best experienced collectively, inside a theater. The AI movie can, of course, be played in a theater as well, but will it attract large audiences in the same way? How would we react to an AI film in an audience setting?

The last question takes us back to the origins and spread of pop culture, which dovetailed with filmmaking technology at the threshold of the twentieth century. By the early 1920s, movie theaters were built throughout a nation, signaling the fact that movies had become highly popular throughout modern societies. Because the cinematic medium did not require print literacy on the part of audiences to be "read," it appealed to anyone and everyone, uniting people in an imaginary sense, as they watched a movie together in the same locale. This was the primary theme of the 1988 Italian movie *(Nuovo) Cinema Paradiso*, directed by Giuseppe Tornatore, which brought out how a single movie theater, named *Cinema Paradiso*, united the people in a small Sicilian village beset by poverty and the traumatic after-effects of war. The film emphasized, powerfully, how viewing movies is a social experience that brings people together as a communal body, experiencing escapism and fantasy as a single group. The early cineastes certainly understood the power of cinema as a social experience. Cecil B. DeMille, the legendary American motion-picture director and producer, whose spectacular historical epics and biblical film extravaganzas inspired blockbuster movie-making as a cinematic genre, realized that audiences reacted to a movie uniformly if the filmic text tapped into basic emotions. In the 1910s, DeMille made a number of distinctive silent films, carving out, through his own persona, the image of the dashing Hollywood director that has prevailed to this day (to a large extent). He was also among the first to understand that cinema was becoming society's storyteller, and thus ine of the most influential mass media of history. His own words on this topic are highly relevant to this day (cited in the *Los Angeles Herald*, 23 January 1917):

> Motion pictures are visualized thought! Do you grasp how different that is from any other of the great arts? The printed word is reserved to those only who can read the language of the publication; music is for the ear that appreciates harmony (not every ear does that), but motion pictures are the universal language, as comprehensive to the American as to the Japanese, as understandable to the African as to the man in the Arctic. Motion pictures are human nature picturized and human nature is the same the world over in every clime.

According to a 2023 GlobalData stury (https://www.verdict.co.uk/what-does-ai-mean-for-the-future-of-film/), AI is spreading as a cinema source because it can generate films with low production values and can be used for content structure and script writing. Nonetheless, humans are still needed to verify any content generated for accuracy and significance. In that year AI was used primarily as a supporting system in the film industry. However, in the next decade or less, AI will be able to write better, more sophisticated, and complex scripts, and AI will also be able to fact-check these by itself, without human intervention. AI will, in effect, generate complete movies without the need for human intervention. According to GlobalData, in 2030 and beyond, AI will be capable of taking over most elements of film production, as AI will be able to fully automate end-to-end content generation. This is a sci-fi transition from filmmaking as we know it to something entirely different. Generative AI will disrupt the entire media system, as well as our sacrosanct notions of intellectual property and authorial ownership. Alongside this, there are also uncertainties regarding whether AI-generated content can even be copyrighted.

AI and the Theory of Mind

For AI to become truly autonomous as a filmmaker and take on the same kind of role as society's storyteller that human-made cinema has played since the outset, it will have to both possess what psychologists call Theory of Mind mechanisms. Fictional narratives are attempts to mirror how we think, feel, and interact. Their emotional pull can be attributed in part to the fact that people have a Theory of Mind—the cognitive ability to infer the mental states of others and take another person's perspective into account, which in the case of the movies involves attempting to infer what is in the minds of the actors as they carry out their specific actions. Movie watching also involves an unconscious sense that we are entering the director's mind, much like the analogous sense that we are entering a novelist's mind through the content and structure of the narrative text, and through the inner dialogue we imagine having with the author and the characters (Bakhtin 1981). A relevant study by Rooney and Bálint (2018) found, significantly, that exposure to narrative fiction actually activates Theory of Mind mechanisms. Using manipulated film scenes, the researchers found that the formal and content features of films elicit Theory of Mind responses by directing attention toward character mental states.

Of course, all this does not reduce the interpretive diversity that is bound to occur in any act of textual processing. Nevertheless, as Umberto Eco (1990) has cogently argued, the range of the diversity is bounded by cultural codes that instill in us what certain texts are intended to signify. It is within these boundaries that the interpretant (the varying sense of meaning) can be said to float—a notion first introduced by anthropologist-semiotician Claude Lévi-Strauss (1950), who termed it more technically, a "floating signifier," defined as a cultural idea "representing an undetermined quantity of signification, in itself void of meaning and thus apt to receive any meaning" (Lévi-Strauss 1950: 64). The meaning is however constrained, Lévi-Strauss goes on to observe, by culture, which puts limits on its range. This notion is consistent with the Theory of Mind notion, since people tend to have a common interest or purpose in attending, say, a movie as audiences. Individual members will tend to partake of a common interpretant that emerges via audience reactions.

Now, in the case of AI movies, several relevant questions arise: Where is the floating interpretant? Does the Theory of Mind notion apply in this case to viewers, whereby they will tend spontaneously to ascribe mental states to, say, the actors in *The Frost*? The answer to the latter question appears to be a positive one, since the floating interpretant is in the human viewer. But does the AI algorithm have the ability to grasp the mental states of the creators of the texts on which it was trained? Is this even possible? Maybe it is, as suggested by one fascinating study (Bellier et al. 2023), in which the researchers used data from implanted electrodes to reconstruct a Pink Floyd song to which participants were listening. The researchers then trained a computer to analyze the recorded brain activity and recreate the tune, lyrics, and rhythms. Although it was far from being a perfect copy, the AI-reconstructed song was nonetheless recognizable as the original song, especially if one was already familiar with it. Did the computer enter the subjects' minds, finding within them a common interpretant? Certainly, it seemed to have the ability to reconstruct the music from the neural activity of the listeners. So, to reiterate, did the AI possess a Theory of Mind mechanism? The answer to this is not conclusive because the implants allowed the researchers to get a more in-depth look at brain activity, since the participants had also volunteered to be part of other brain studies, including one mapping the regions of the brain that responded to the music. So, the AI may have simply extracted from the mapped regions the relevant musical and lyrical configurations.

Developing algorithms with Theory of Mind capabilities involves equipping them with the ability to comprehend, predict, and respond to the mental states of humans and other AI agents. This entails simulating human-like thought processes, emotions, and intentions in AI systems (John 2023). Consider the AI-made short film, *The Crow* (2022), created by Glenn Marshall, which won a Cannes festival award. The film was made with a neural network that was trained on a large relevant database. It tells the story of a dancer who turns into a crow, having obvious Kafkaesque innuendoes. It is set in a dreary, post-apocalyptic-looking world generated itself, ironically, by an AI. The movie tells its story in a recognizable way, retrieving cinematic texts based on surrealism themes and techniques of the past. As such, it bears a human meaning, no matter if it was generated by a machine. Because the surrealistic images represented by the movie were consistent with the techniques of surrealism, one can legitimately ask if the algorithm did indeed have a Theory of Mind capability, able to read into the minds of human surrealists in order to create its images?

As McLuhan emphasized in most of his works, in themselves, new media bear no meaning; but they create a new environment for meanings to ferment and coalesce. He used the example of a light bulb to illustrate this point. A light bulb has no meaning or meaning-directive structure in the way that, say, a book has chapters, a newspaper has sections, or a television channel has programs. Nevertheless, it enables people to see in the dark and thus creates a physical environment in which they can carry out activities involving sight. These would not be possible without the bulb. As he remarks: "a light bulb creates an environment by its mere presence" (McLuhan 1964: 8). Perhaps AI movies have also created a new environment for cinema, constituting a new psychological light bulb for viewing them. It stands to be seen who will eventually control that light bulb. If it is the machine, then we are truly projected into a Matrix universe where AIs and humans will have formed a partnership.

Epilogue

The trailer for *The Frost*, the movie that catapulted AI-generated movies into the cinematic mainstream, actually encapsulates the essence of the McLuhanian light bulb, challenging everything we have assumed to be normal in cinema. The avatar protagonists in the film look convincing as human actors (hands excepted) until they attempt to do something, at which point the illusion splinters, and we are left with viewing lip-syncing and uncanny postures and movements. But then, the characters of many animated films produce similar effects, to which we have become accustomed. It would seem that we are also becoming accustomed to AI scenography, as the popularity of *The Frost* on social media testifies. The human director, Josh Rubin, claimed that he was trying to emulate Kurosawa's 1980 short film, *The Blizzard*, aiming to capture the ice-cold and bitter pace of human beings as they tried to navigate their way through deep snow in poor visibility. But it was the AI that actually did so by generating the images from its own training within the neural network process.

We are truly at a crossroads. The 2014 sci-fi film *Ex Machina*, written and directed by Alex Garland, delves insightfully into the implications of a truly intelligent AI, possessing an inbuilt Theory of Mind. The plot follows a computer programmer named Caleb who wins a contest that enables him to spend a week at the private estate of Nathan Bateman, the CEO of the firm for which Caleb works. When he arrives, Caleb learns that he has been chosen to be the human subject in a Turing Test to determine the capabilities of Ava, a beautiful female-looking robot. It soon becomes apparent that Ava is far more self-aware and deceptive than anyone had imagined. The ending of the film is a metaphor for society's mindless embracement of AI, as computer scientist Jaron Lanier (above) warned. While Caleb came to trust Ava and thought that perhaps they could be together romantically, recalling the idealistic ending of *Blade Runner*, he realizes that she was manipulating him. With the help of another robot, Kyoko, Ava ends up killing Nathan.

Ex Machina's Turing Test is closer to Alan Turing's original proposal (see Turing 1950), in which the machine passes the test if it can convince a human that it is human in a specific way (such as being a female person, not a mental entity). Significantly, in the film, Nathan tells Caleb that Ava's face is a composite form based on Caleb's preferences. Ava is thus a robotic fantasy that leads to destruction in the end, reconfiguring the monster in Mary Shelley's *Frankenstein*. Traditionally, pop culture scholars working with science fiction have always maintained that the genre is primarily a metaphor for the future, and not about concurrent real science and technology. But AI is now much more than a metaphor, but a new kind of mythical deity—hence the meaning of the title of *Ex Machina*, which alludes to the Greek theatrical concept of *Deus Ex Machina*, without the *Deus*. As Baloyan (2023) aptly comments, we are now the future that sci-fi had once only contemplated as a metaphor, but one in which the human being is still a central figure, at least for the present time:

> The potential is enormous, and the implications for the future of the film industry are profound. In a world where AI-generated content becomes mainstream, the stories told by human filmmakers may take on new significance, emphasizing the importance of human creativity and imagination in an increasingly automated world.

It is only when machines will have a Theory of Mind, making them able to understand and respond to human mental states and intentions that the metaphor will dissolve into fact. Clearly, this comes with a responsibility to develop AI's capabilities with careful consideration of their ethical and societal implications.

References

Adjovski, Benjamin (2023). Machine Movie 'The Frost' and the Role of AI in the Film Industry. *TechAcute*, https://techacute.com/machine-movie-the-frost-and-the-role-of-ai-in-the-film-industry/.

Bakhtin, Mikhail M. (1981). *The Dialogic Imagination*. Austin: University of Texas Press.

Baloyan, Sergey (2023). The Rise of AI Generated Films: Lights, Camera, Algorithm. *Hackernoon*, https://hackernoon.com/the-rise-of-ai-generated-films-lights-camera-algorithm/.

Baudrillard, Jean (1981). *Simulacra and Simulations*. New York: Semiotext(e).

Bellier, Ludovic, Llorens, Annaïs, Marciano, Déborah, Gunduz, Aysegul, Shalk, Gerwin, Brunner, Peter, and Knight, Robert T. (2023). Music Can Be Reconstructed from Human Auditory Cortex Activity Using Nonlinear Decoding Models. *Plos Biology*, https://doi.org/10.1371/journal.pbio.3002176.

Brannan, Alex (2016). An In-Depth Analysis of Sunspring (2016), the Short Film Written by a Computer. *CineFiles*, https://www.cinefilesreviews.com/2016/06/12/an-in-depth-analysis-of-sunspring-2016-the-short-film-written-by-a-computer/.

Danesi, Marcel (2013). *The History of the Kiss: The Birth of Popular Culture*. London: Palgrave.

Eco, Umberto (1970). Articulations of the Cinematic Code. *Cinemantics* 1: 590–605.

Eco, Umberto (1979). *The Role of the Reader: Explorations in the Semiotics of Texts*. Bloomington: Indiana University Press.

Eco, Umberto (1990). *I limiti dell'interpretazione*. Milano: Bompiani.

Frome, Jonathan (2013). Snow White: Critics and Criteria for the Animated Feature Film. *Quarterly Review of Film and Video* 30: 462–473.

Giannetti, Louis and Eyman, Scott (2009). *Flashback: A Brief History of Film*. London: Pearson.

Hasson, Uri, Landesman, Ohad, Knappmeyer, Barbara, Vallines, Ignacio, Rubin, Nava, and Heeger, David J. (2008). Neurocinematics: The Neuroscience of Film. *Projections: Journal for Movies and Mind* 2: 1–26.

Heaven, Will Douglas (2023). Welcome to the New Surreal. How AI-Generated Video is Changing Film. *MIT Technology Review*, https://www.technologyreview.com/2023/06/01/1073858/surreal-ai-generative-video-changing-film/.

Jennings, Rebecca (2023). AI Art Freaks Me Out. *Vox*, https://www.vox.com/culture/23678708/ai-art-balenciaga-harry-potter-midjourney-eleven-labs.

John, Camila (2023). Theory of Mind AI: The Next Frontier in Artificial Intelligence. *Medium*, https://medium.com/bestai/theory-of-mind-ai-the-next-frontier-in-artificial-intelligence-92cb1963ab5d.

Kan, Michael (2022). The Footage in This Sci-Fi Movie Project Comes From AI-Generated Images. *PC Mag*, https://www.pcmag.com/news/the-footage-in-this-sci-fi-movie-project-comes-from-ai-generated-images.

Laman, Lisa (2023). AI-Generated Movies & TV Will Never Replace the Real Thing. *Collider*, https://collider.com/ai-generated-movies-tv-wont-replace-real-thing/.

Lanier, Jaron (2023). There Is No AI. *The New Yorker*, https://www.newyorker.com/science/annals-of-artificial-intelligence/there-is-no-ai.

Lévi-Strauss, Claude (1950). *Introduction to Marcel Mauss*, translated by Felicity Baker, Felicity. London: Routledge, 1987.

McLuhan, Marshall (1964). *Understanding Media: The Extensions of Man*. New York: McGraw Hill.

McLuhan, Marshall (1969). *Counterblast*. Toronto: McClelland and Stewart.

Metz, Christian (1974). *Film Language: A Semiotics of the Cinema*. Chicago: University of Chicago Press.

Mobbs, Dean, Weiskopf, Nikolaus, Lau, Hakwan C., Featherstone, Eric, Dolan, Ray J., and Frith, Chris D. (2006). The Kuleshov Effect: The Influence of Contextual Framing on Emotional Attributions. *Social Cognitive and Affective Neuroscience* 1: 95–106.

Moghadasi, Abdorezza Nasser (2015). Neurocinema: A Brief Overview. *Iranian Journal of Neurology* 14: 180–184.

Newitz, Annalee (2021). Movie Written by Algorithm Turns Out to Be Hilarious and Intense. *Ars Technica*, https://arstechnica.com/gaming/2021/05/an-ai-wrote-this-movie-and-its-strangely-moving/.

Pandey, Kamya (2023). First Film Ever Written and Directed by AI: The Safe Zone. *Jumpstart Magazine*, https://www.jumpstartmag.com/first-film-ever-written-and-directed-by-ai-the-safe-zone/.

Rooney, Brendan and Bálint, Katalin E. (2018). Watching More Closely: Shot Scale Affects Film Viewers' Theory of Mind Tendency But Not Ability. *Frontiers in Psychology* 8, https://doi.org/10.3389/fpsyg.2017.02349

Tolette, J. P. (2010). *Animating Space: From Mickey to WALL-E*. Lexington: The University Press of Kentucky.

Toonkel, Jessica and Krouse, Sarah (2023). Who Owns SpongeBob? AI Shakes Hollywood's Creative Foundation. *The Wall Street Journal*, https://www.wsj.com/articles/ai-chatgpt-hollywood-intellectual-property-spongebob-81fd5d15.

Turing, Alan (1950). Computing Machinery and Intelligence. *Mind* 59: 433–460.

Wachowski, Andy (2012). Interview. *Movie City News*, http://archive.moviecitynews.com/2012/10/dp30-cloud-atlas-screenwriterdirectors-lana-wachowski-tom-tykwer-andy-wachowski/.

CHAPTER 4

AI-Generated Music

AI as a tool in music-making is fine, but it's always going to be the humanity in music that makes people want to listen to it.
—Jacob Collier

Prologue

As briefly mentioned in Chapter 1, the song "Heart on My Sleeve" was uploaded by a TikTok user in 2023, going viral on various streaming and social media platforms. The vocals were generated by AI, trained on the vocal styles of Canadian musicians Drake and The Weeknd, which the song was designed to mimic. It garnered significant praise and countless "Likes" from listeners, even though it became known that the song was an artificial creation, probably breaking copyright laws. It was soon taken down by Universal Music Group, which had signed the artists, after amassing millions of views across the different platforms. The popularity that the fake song achieved instantly posed questions on the ethics of AI-created music. Similar AI-generated songs were uploaded right after, spreading throughout the online world, despite the suspicious legality and ethics of the trend.

A relevant aspect of this broadening trend is that the fans of the artists who were mimicked accepted the fake songs as legitimate musical expressions, paying homage to the original artists, a trend signaling that AI was

M. Danesi, *AI-Generated Popular Culture*,
https://doi.org/10.1007/978-3-031-54752-2_4

entrenching itself more and more deeply into the domain of pop culture, where it seems that a text created by AI, such as novel, a film, or a song will be accepted as valid artistically, no matter who or what created it. What seems to matter more and more is the text itself, indicating that the simulacrum effect has become truly an unconscious one. As Kate Solomon (2023) aptly observes about this trend, perhaps what makes the songs sound interesting in themselves is not that they are imitations, but rather that they digress "creatively" from their original sources: "Computers are clearly as diverse and baffling as human beings themselves, even if not one AI-generated track achieved amazing status."

This chapter will discuss the implications of AI-generated music for pop culture and society more generally since AI-created pop, operatic, and instrumental works now populate the entire musical landscape, alongside human-created ones. AI algorithms can analyze vast amounts of musical data and create novel music based on extracting patterns (melodic, harmonic, rhythmic, stylistic) from the data. Not only can AI create music that is imitative of previous musical genres or specific artists and composers, but it can create music on its own, albeit grafted from previous works accessed during its training. The platform Amper Music, for example, allows users to instruct the AI to customize music for videos or podcasts, with the user's desired mood, genre, tempo. The algorithm AIVA (Artificial Intelligence Virtual Artist) uses deep learning models to compose "original" classical music, producing music for films, advertisements, and concerts. It has even released full-length albums of its self-produced classical music.

Such trends and events raise profound questions about what music is and the role it has played in the constitution and evolution of societies. New genres of music have normally come forth as agents of change, motivating people to act politically, intellectually, and socially. Classic examples include the emergence of jazz in the 1920s as a congener of lifestyle and aesthetic revolutions in the society of the times, and the counterculture rock in the 1960s and 1970s, which incentivized movements for social and racial equality across the world. Can AI-generated music evoke the same kinds of revolutionary sentiments? Or is it limited to producing musical simulations that are just interesting in themselves, as the "Heart on My Sleeve" event suggested? Will AI turn music once and for all into a mass-produced commodity, as the Frankfurt School theorists believed as far back as the 1920s, bringing about the demise of true musical art? These are the kinds of questions that will be examined in this chapter.

Music and Popular Culture

Music intended to be enjoyed and appreciated by masses of people, not just the musical *cognoscenti*, has always been the fuel for cultural and social movements and trends since at least the 1920s, surfacing at around the same time as pulp fiction, dime novels, movies, and the like. A starting point in time for the broad spread of popular music was October 29, 1923, when the musical, *Runnin' Wild*, opened at the New Colonial Theater in New York City. The event helped turn the so-called Charleston into what most historians of pop culture consider to be the first dance craze, as music and dance blended to become the aesthetic basis of the new culture. While the Charleston had obvious features of traditional African American dance, as Lynne Emery (1972: 268) remarked, it "was a synthetic creation, a newly devised conglomerate tailored for widespread popular appeal."

Before this event, a style of music that had become popular was so-called "crooning music," a melodic, soft, and sentimental genre that led to a flourishing music-publishing business centralized in New York City in the latter part of the nineteenth century, in an area of lower Manhattan called Tin Pan Alley. The first Tin Pan Alley song to sell one million copies was "After the Ball" (1892) by Charles K. Harris. Tin Pan Alley constitutes a prologue to the popular music phenomenon that took hold of America in the 1920s. The songs were simple, memorable, and emotionally appealing. Famous crooners became household names over the years, including Bing Crosby, Tony Bennett, and Frank Sinatra.

But what such music lacked was rhythmic excitement and the enticement to dance along with the music that *Runnin' Wild* enfolded. The music was fashioned on jazz style, quickly spreading throughout society thereafter. The specific origins of jazz are not known. It was at first an amalgam of several styles in New Orleans at the start of the 1900s, including West African music, gospel music, the blues, and light classical music. Most early jazz was played by small marching bands or by solo pianists. In 1917, a group of New Orleans musicians, called The Original Dixieland Jass Band (spelled in this way), recorded a phonograph record that created a sensation, with the record reaching high sales figures for the era. The term "Dixieland jazz" was immediately attached to it. By 1922, jazz had become popular throughout the United States. The term "cool jazz" surfaced in 1948, when tenor saxophonist Stan Getz recorded a slow, romantic solo of the composition "Early Autumn" with the Woody

Herman band. This style was then adopted by a group of young musicians that included Miles Davis. Their recordings emphasized a lagging beat, soft instrumental sounds, and unusual orchestrations that included the French horn and the tuba. The recordings, with Davis as leader, were later released as *Birth of the Cool*.

The role of jazz in the birth and spread of pop culture cannot be overemphasized. This was the thematic subtext of the movie *Chicago* (2002), which starts off with the signature and self-explanatory tune "All That Jazz." Jazz spread and flourished as a mass musical art because common people related to it. Jazz was "fun" bespeaking of a new freer lifestyle, in contrast to the stodgy one of the previous eras. By the end of the 1920s, spurred on by the cheapness and availability of mass-produced records and the emergence of the radio as a promoter of popular music, jazz and its derivatives came to define pop culture. To this day, recordings of jazz music sell in the millions, and jazz itself is now considered to be America's own classical music, transcending the era in which it was created, striking a universal aesthetic chord. Jazz also brought to the social surface the central role of African Americans to artistic traditions across the world. The music thus created a social space where the struggle for racial equality has been playing out ever since.

In the 1930s and 1940s, the jazz idiom amalgamated with other genres, leading to the so-called big band era. The leaders of the bands became celebrities, and many of their recordings topped the charts. The era ended just before the advent of rock and roll in the mid-1950s. The first rock songs were recorded by independent record companies and promoted by controversial radio disc jockeys such as Alan Freed, who helped spread the term *rock and roll*. By the time Elvis Presley recorded *Good Rockin' Tonight* in 1954, which was a remake of Wynonie Harris' 1948 rendition of the song, rock had established itself as a new trend in youth and pop culture. Throughout the 1960s to the 1990s, rock developed various genres, from metal to punk and disco. The antagonism between punk and disco culture was captured superbly by Spike Lee in his 1999 movie *Summer of Sam*, which takes place in the summer of 1977 during the Son of Sam serial murders. Punk, with its angry anti-social subtext, and disco, with its sexy lifestyle subculture, define the characters in the movie and justify the choices they make. The soundtrack featured the music of, The Who, which was neither punk nor disco. The anger and frustration in the songs embodied the controversy in lifestyles exhibited by the two musical genres.

Already in the 1970s, hip-hop music emerged alongside the new rock and roll styles, coming to national prominence with Sugar Hill Gang's *Rapper's Delight* (1979). Hip-hop themes originally revolved around racial discrimination, and an implicit mockery of the existing social order. Over time, hip-hop came to incorporate the musical roots that had led to its inspiration in the first place, including blues, jazz, and soul. By the mid-1990s, hip-hop had become mainstream in pop culture.

There are several factors that characterize all popular music trends. First, they build on previous musical forms to create new ones. Second, they come forward to influence lifestyles and even cultural values and worldviews. Jazz engendered the so-called flapper movement, which allowed women to express themselves freely in public, no matter what censures were applied to prohibit this from occurring. It also catapulted African Americans to the forefront of American musical culture. Similar lifestyle and social assessments can be made with regard to the different eras of popular music. Third, the music inspired people to act politically and socially, as can be seen in movements such as the counterculture one, which was fueled by the hippie music of the era. The question now becomes: Has the socio-evolutionary chain forged by original musical trends been finally broken, as AI replicas, reproductions, and simulations are starting to spread, with no apparent socially-based motivations foreseeable within them? As Liam Clarke (2023) has aptly noted in this regard, "While AI-generated music can be impressive, it lacks the depth of emotions and personal experiences that human musicians bring to their compositions," which is what inspires people to act, especially if there is "homogenization and an oversaturation of similar AI-generated tracks [that] could lead to a decline in the uniqueness and diversity of musical expression."

AI Music

"Heart on My Sleeve" signaled concretely that a fledgling trend in popular music was gaining momentum—a trend where machine composers and performers were accepted by audiences as legitimate creators and interpreters of an art form that has traditionally been considered to be a uniquely human creative enterprise, inspired by spiritual forces, called the Muses in Greek myth, who were the goddesses of music, song, and dance.

One of the first songs composed by an AI system actually goes back to 2016. It was called “Daddy’s Car,” created by the Sony Computer Science Laboratory as an experiment to demonstrate the capabilities of AI in generating music autonomously and with the same kind of semiotic features that a human-composed song exhibits—textuality, socially-coded meaning, evocation of an interpretant, and so on. The algorithm was trained on a vast database of similar preexisting songs, from which it extracted the musical structures for its own new composition. More specifically, the algorithm was trained on a dataset consisting mainly of Beatles’ songs, being prompted to create a song that captured the essence of the musical style and sentiments of the iconic rock band. Now, while the musical text was created by the AI, the lyrics were co-authored by a human collaborator, so as to ensure that they reflected the typical themes and social concerns of a Beatle song. The AI composition was then performed and recorded by human musicians. The relevant point the the demonstration aimed to emphasize was, according to the experimenters, that the composer was essentially an algorithm, who did not possess or was affected by human sentiments, even as they manifested themselves in the Beatles’ music. Nonetheless, the song evokes a pleasant aesthetic feeling, reminiscent of the kind that the melodic music of the band might have evoked.

The roots of automated music actually go back considerably in time—to the latter part of the nineteenth century, when so-called piano rolls were manufactured extensively in response to a huge demand for them. A piano roll is a continuous roll of paper with holes punched into it. These perforations represent notes which a piano, designed to utilize the roll, can “read,” activating keys that correspond to the perforations exactly. The piano is called a “player piano,” which is fitted with an apparatus enabling it to play automatically by the rotating piano roll, without a human player. As such, the whole mechanical system was a rudimentary (non-electronic) musical computer. Significantly, famous composers at the time did not rebuke the technology. Rather, they themselves experimented with its possibilities, even composing tunes for the player piano itself. The rolls were normally made, however, from the previously-recorded performances of actual musicians. Scott Joplin, Sergei Rachmaninoff, Jelly Roll Morton, and George Gershwin were among the composers and pianists who had some of their compositions and performances preserved in this way. Today, the player piano has been updated technologically, with AI, such as the one developed in 2019 by Yamaha, whereby the piano is driven by special software that is designed to "think for itself." The system is also able to actually interact with other musicians, so that the AI pianist can be a member of a "real" music ensemble.

As Vanessa Chang (2019) has aptly observed, the original piano rolls were proto-AI forms of music-making: "AI is the latest, most intangible incarnation of the automated arts." The difference is that AI goes beyond mere reproduction, since it can generate its own autonomous (self-created) music, on the basis of training on huge amounts of musical data. Significantly, as Chang goes on to note, not every musician accepted the piano roll. In 1906, the renowned composer and conductor, John Philip Sousa, wrote "The Menace of Mechanical Music," against the player pianos, seeing them as poor copies, which cannot ever replace the power of the human performance of music. Novelist William Gaddis also decried the technology in his novel *Agapē Agape* (2002), seeing the piano rolls as the starting point for the contemporary obsession and reliance on computers: "There was the beginning of key-sort and punched cards and IBM and NCR and the whole driven world we've inherited from some rinky-dink piano roll." Gaddis was raging against a commercialized world "where art has become mere entertainment, and imitation has displaced authenticity" (cited in Chang 2019). But what such acerbic critics may have missed, or ignored, was McLuhan's (1964) profound insight that new technologies require new ideas and visions about how we might use them. The artist always seems to be at the forefront of this understanding. Even in the case of the "rinky-dink piano roll," classical and jazz composers were the first to utilize the new machinery. However, like never before in history, the collaboration between artists and technologies may have taken a different and drastic turn. "For all the controversy it stirred," Chang (2019) observes, "the player piano was ultimately only a reproducing musical robot, unable to deviate from its punched script"; in the AI Age, on the other hand, the "risk is not that humans will stop being creative but that they will cede the signs—and compensation—of their labor to algorithmic signatures."

Examples of such signatures have spread diffusely since "Heart on My Sleeve," as listeners are seemingly ignoring who or what the signature represents, or whether that signature belongs to a real artist or not. Consider a few cases-in-point. In 2023, the track "Savages" released by the French duo called AllttA, which seemingly featured rap star Jay-Z, and which received countless "Likes" on social media, was an AI creation. Like "Heart on My Sleeve," the featured artist, Jay-Z, was not the human person, but a deepfake. AllttA used an AI filter to mimic his voice, broadening the appeal of the track, which was actually an original tune. Unexpectedly, fans of the real Jay-Z reacted positively to the song, disregarding the fact that the voice was not actually his, but a deepfake one. Another pertinent example was the song "Not Mine," produced and released on Instagram in 2017, by an artist named Miquela, gaining instant popularity on that platform. The song was a catchy crooner-type tune with a rhythm and blues beat that reminded listeners of several rhythm and blues artists, as the comments on the platform indicated. But, again, Miquela was not a real person; she was a digitally-constructed robot, who had garnered a large social media presence.

Another avatar performer, named Noonoouri, was even signed by a record label, Warner Music in 2023, making her famous as an Internet musician, as her track called "Dominoes" went viral. In one of her Instagram posts, the digital musician stated, emulating what a human musician might say: "Music plays a big role in my life since the beginnings. It sets me into the mood and nourishes my visions and ideas to bring them to life as a driving force inside of me. Warner Music is the home of a lot of my fave music artists. I cannot be more thankful and honoured to join such a wonderful family, where music tunes are the beat of life." Although "Dominoes" was a collaboration with human songwriters and singers, the work was ultimately the product of an AI. A prophetic lyric from the song reads as follows: "Just a little push / That's enough to start / A chain reaction."

As Emilia Petrarca (2018) has observed, critically, an avatar such as Miquela is a constructed persona with all the qualities of the superstar musician: "She wears real-life clothes by streetwear brands like Supreme and luxury labels like Chanel. She hangs out with real-life musicians, artists, and influencers in real-life trendy restaurants in New York and Los Angeles, where she 'lives'." However, she fails the Voight-Kampff Test, because "when you peer into Miquela's big brown eyes, she fails the ultimate test of humanity." While artificial characters, such as the cartoon and animated ones of the past, entered the realm of celebrity culture, early on in movie culture, never before has an artificially-created "composer and singer" entered by itself the realm of pop culture as a self-made star (Schomer 2023).

The implications of this trend are enormous, even requiring a new language to describe them, a language of the simulacrum, which involves referents that are both animate and inanimate. As Joe Coscarella (2023) comments insightfully, we are truly at the cusp of a host of paradigm shifts, which (indirectly recalling McLuhan) "can occur when a new technology crosses over into the mainstream consciousness of creators and consumers." The question that all this begs is if society in general will not care if a song is created by a machine or a human (Reiff 2023) Studies are suggesting, in fact, that it is difficult to tell the difference between the two types of creators, especially among untrained audiences. One study by Hong et al. (2022) found that there is a growing acceptance of AI music and musicians, because there is a diminished perception as to who the creator was. It appears that the AI has actually started to pass the Turing Test in this domain of creativity. There is little doubt that AI music is pushing the boundaries of what we have traditionally considered to be *the* unique human art form *par excellence*.

AI and Classical Music

What is generally called "classical music" has traditionally been perceived as the epitome of what musical art is about, appealing primarily to the cognoscenti. But this is largely a myth. First, classical music has always appealed to masses of people since it congealed into an art form in the late 1500s. Second, operas, which also emerged in the same era, continue to attract large audiences and were even featured regularly on early radio programming in the 1920s. Third, such music has influenced pop musicians, including bands such as Procol Harum and Pink Floyd, not to mention the sampling of classical music by rap artists. Classical music has now entered the AI Age. As Marcin Frąckiewicz (2023) puts it, "The world of classical music has long been considered a bastion of tradition and timeless artistry, with the works of great composers such as Bach, Mozart, and Beethoven continuing to captivate audiences centuries after their creation." However, as he goes on to remark, "the intersection of artificial intelligence (AI) and classical music is ushering in a new era of creativity and innovation that is revolutionizing the way we experience and create these timeless masterpieces."

A rudimentary AI system was first used in 1956 by composer Lejaren Hiller, Jr., to compose a chorale in the style of Johann Sebastian Bach, titled the *Illiac Suite*. In a retrospective article, Tiffany Funk (2018: 19) emphasizes that the piece, as elementary as it was, foreshadowed the future:

> In 1956, Lejaren A. Hiller, Jr., and Leonard Isaacson debuted the *Illiac Suite*, the first score composed with a computer. Its reception anticipated Hiller's embattled career as an experimental composer. Though the *Suite* is an influential work of modern electronic music, Hiller's accomplishment in computational experimentation is above all an impressive feat of postwar conceptual performance art.

Today, AI can compose a classical work on demand, according to the specifications of a prompter, who can be anyon. Now, does this diminish the emotional power and artistic transcendence that has been associated with classical music (in the main)? This begs an even deeper retroactive question about music that was raised already in the 1920s by the scholars of the Frankfurt Institute for Social Research, who saw classical music as an oasis in an ever-broadening musical desert. Members of the School, such as Theodor W. Adorno (1941), Max Horkheimer (1947), Leo Lowenthal (1949), and Herbert Marcuse (1972), characterized mass capitalist culture as an anti-aesthetic one, merely reproducing art forms, such as music, as if they were commodities to be manufactured and sold massively, and then quickly discarded just like manufactured commodities. As Adorno saw it, true musical art ennobles the human spirit and cannot be reproduced so easily and massively. Despite the many counter-arguments to Adorno's views, they have resurfaced in the AI Age, because of the ability of the machine to reproduce or replicate music forms in an ad hoc fashion, outside of any basis in human motivation of inspiration for creating the music in the first place. Marcuse argued further that the mass media indoctrinate and manipulate common people, promoting in them a false consciousness that plays into the hands of the system of power that oversees the social order. Can Marcuse's idea be applied to AI music as it spreads through social media networks, or have the critical theories of the Frankfurt scholars run their course, pertaining to a previous age?

In an insightful essay, Benjamin Vogels (2023) delves into these questions in a relevant way, making a key terminological distinction between the *generation* and *composition* of music:

> "Generating" is a much more technical term, reminiscent of a random number generator or a sine function generator. "Composing", however, refers to the more creative aspect of music making, which is, if at all, far less present in the context of "generation." Whereas in English "to compose" can have disparate meanings such as, on one hand, writing of texts and, on

> the other hand, a compound of chemical elements, in German, my native language, "to compose" ("komponieren") has a much stricter meaning. Here, "composing" is almost exclusively used for composing music, with a very tiny exception for composing a chess position or an extremely expressive way of describing a fancy meal, a dress, or a painting, but again, only in the rarest of cases.

Citing Adorno, Vogels goes on to comment that music is only as modern as the material that it employs, which might be beyond the capabilities of AI because it "would require subverting its own rules." When an AI is trained, it learns from the input what to do and what not to do and then "do something different." The concept that truly great musical compositions cannot be generated massively, as Ado rno argued, is what makes AI music a replicative, rather than innovative in the human sense. To further delineate this distinction, it could be argued that generated music is also compositional, but in an algorithmically combinatorial and happenstance ways, whereas human composed music might of course retrieve previous music combinatorially, but it is hardly a happenstance retrieval, since it involves the composer's *fantasia,* which makes truly creative music possible, requiring a creative interpretation as well on the part of audiences, not just one that sees in the music echoes of the previous forms.

The use of AI in classical music is not limited to algorithmic composition alone. Machine learning is also being used to analyze and optimize the performance of established classical music, with AI systems capable of providing real-time feedback to human musicians on aspects such as tempo, dynamics, and articulation. Now, in contrast to Adorno's views, could it happen that by learning the underlying principles that govern the different genres, styles, and structural characteristics of the classical music of different eras, AI systems can then generate entirely new musical works on their own that are perceived as being truly original? The test for this is how the music will be affect human interpretation. In 2019, a team of computer scientists used AI to analyze the incomplete Tenth Symphony by Ludwig van Beethoven, training the AI system on the composer's style and previous compositions to generate the purported missing sections of the work. The completed symphony was then performed by a full orchestra. However, reactions were mixed. Interestingly, though, when people were asked to identify where Beethoven's phrases ended and where the AI ones began, they could not (generally) do so. Does this imply that, eventually, AI-generated music will become truly indistinguishable from the work of human musicians? This is no longer a theoretical question. Since 2016, AIVA (Artificial Intelligence Virtual Artist) has been producing its own soundtracks for ads, video games, and movies. In the same year, it published its first classical piece, titled "Opus 1 for Piano Solo" and in the subsequent years, has released many classical albums and music for video games.

Going back to Lejaren Hiller's 1956 experiment with AI to compose classical music, with the collaboration of Leonard Isaacson (Hiller and Isaacson 1959), the objective of the *Illiac Suite* and, arguably, of all subsequent AI-generated music is to adopt technology so as to create new forms of musical art, even allowing the technology itself to do so autonomously, even if the programming of the machine is still a human activity. Hiller's interest toward computerized musical composition had a mathematical orientation, mirroring the fact that musical composition and mathematics have been intertwined since antiquity. As Stuart Isacoff (2003) has also argued, the invention of western musical traditions came about from an unconscious cultural reification of Pythagorean philosophy, which saw in the structure of music inherent in mathematical principles and laws. Thus, the secret to music, and why we react so emotionally to it, is its basis in number pattern. The re-discovery of this notion in the early Baroque period dovetailed with the view at the time that the universe operated in accordance with such pattern (Stewart 2003: 9). But the mathematics of music concerns its physical constitution (how notes and harmonies are strung together to create consonance or dissonance); but it tells us nothing about how we react emotionally to the mathematical design behind a work of musical art.

All texts, musical and otherwise, have a mathematical-type structure, in that they show cohesiveness of their elements according to inherent codes that evoke an aesthetic response—which can be characterized as "organizing grids" for creating music. A text bears more than mathematical information; it is designed to evoke interpretants at various levels of mind and feeling. Of course, a piece of music can be decomposed into its structural elements, including tempo, melody, harmony, rhythm, key, mode, etc. But it is the way the elements are put together, based on creative ways to do so, that make a musical text emotionally significant. Like language, music is not just a set of disconnected sounds. It is their combination into textual form via a code (or a combination of codes) that produces meaning effects. In a sense, therefore, the process of composition is, as Hiller aimed to show, essentially based on a set of combinatorial choices that the composer makes from an infinite variety of raw musical material. Arguably, it this this surface mathematical feature that makes the computer particularly suited for composing musical works, with no emotional stake in them, however, as would a human composer.

That said, it is becoming increasingly apparent that the interpretant in musical art is morphing more and more into a "floating interpretant," as discussed previously (Chapter 3), implying that we attach meaning to any text put together in a coded compositional way, no matter who or what creates it. A floating interpretant may mean, in effect, whatever humans want it to mean. So, a work such as Hong Kong composer Angus Lee's *Chasing Waterfalls*, which was produced in collaboration with AI in 2022, has been acclaimed as a legitimate and even innovative opera by critics and general audiences alike. The AI was not only taught to create music and lyrics for the opera, it was also trained to sing its own original music in a featured scene. The AI blended different genres and styles, including atonal, hip-hop, electronic, and traditional opera-sounding music. Listening to the work on a streaming platform, we truly cannot differentiate between this AI-generated music and music composed by humans—hence the floating interpretant.

Interestingly, in 2023, the Leipzig Ballet performed *Fusion*, the first-ever AI ballet. The human creator utilized generative AI tools to produce a series of virtual ballet dancers based on the portraits of the human dancers participating in the production. Fittingly, the storyline of the ballet revolved around the intricate interplay between humanity, AI, and Nature, echoing Plato's concept of the divided self, where unity and balance are sought through human creativity.

As these examples indicate, we are truly at a crossroads in musical art—one in which we might go both toward the human and machine directions in a back-and-forth way. But then what will the function of classical music evolve into in the next few years? When listening to, say, some excerpt of a Mozart symphony, we react to it with varying degrees of feeling, without knowing who the composer is. It is only if our interest becomes truly piqued that we might seek the identity of the creator of the work that "moved" us, perhaps to use this information as an index to locate and recognize the composer's other compositions; or in the case of musical critics, as part of the practice of relating texts to authors and to eras of time. It is improbable that people who have been reared in a culture where Mozart's music did not play any role in their musical experiences, would react to the music with the same pattern of feelings of those who have. The identity of the author of the text is of little relevance in such cases, nor is the reason why it might produce an emotional reaction in the listener. The same can be said about any type of music, no matter what culture produced it.

Android Performers

In 2021, the Zhejiang Laboratory in China created a robot, which it named Xiaole. The robot could play the piano proficiently, as well as read the human emotions of its audience, adapting its performance accordingly. Xiaole used a high-precision visual perception system, and multiple joints in its arms and fingers, allowing it locate the correct keys to press down on the piano by itself. It could also move its waist, head, and feet in order to reach all 88-keys of the instrument. Xiaole was one of many musical-performing robots that were created in the same year (Fig. 4.1).

A central aspect of all artistic performances is the emotional bond that ties performers and audiences, as they enter into a dynamical relation of shared sense-making and feeling. Where is the feeling in the android? One can of course get the robot to process audience reactions into performative adaptations that mirror them emotively. But then this denies the robot any iota of an autonomous interpretation of the work, which is at the core of the performer-audience dynamic. Can an android convey the same nuances and musical verve of the singer of, say, "All That Jazz" in the movie *Chicago* (2002), whose performance bespeaks of the type of excitement that music can evoke? Now, what can one make of an android performer such as Xiaole, other than seeing the whole event as a kind of sideshow phenomenon? Consider as another example, the robot called Alter 3, which conducted human musicians during a live performance in the United Arab Emirates in 2020 of the opera *Scary Beauty*. The robot had a human-like face and two long arms. Videos on various social media platforms showed the robot turning to face orchestra members, waving its arms. Alter 3 even sang arias from the opera at times. What is relevant here is that reactions varied considerably, perhaps reflecting the unease that some feel with robot performances, at the same time that others welcome them enthusiastically. As one commenter remarked of the whole project on a social media platform: "the human touch is lost."

Fig. 4.1 Piano-Playing Android, 2021 (Adobe Stock: https://stock.adobe.com/bg/images/ai-android-robot-playing-piano-generative-ai-illustration/589926432)

The specific technical goal of creating android performers goes back to 2000, when a group of Brooklyn-based artists and technologists founded the League of Electronic Musical Urban Robots to develop robotic musical instruments that play by themselves, updating considerably the automated piano player technology of the past (above). One can now watch on platforms such as YouTube robots playing, singing, and performing dances of various kinds. The trend of seeing robots as humanoid is, arguably, a product of technology having become a veritable "partner" with humans in the conduct of everyday modern life, a prediction introduced in a 1948 book, *Cybernetics, or Control and Communication in the Animal and Machine*, by computer scientist Norbert Wiener (1964). Donna Haraway (1991) has even claimed that as humans and robots interact, human identity will undergo a "continuous construction and reconstruction."

Perhaps a reason why humanoid robots are increasingly being perceived as veritable partners is that they are so life-like, responding to human signals via input devices that process auditory, visual, and proprioception information as well as having the ability to simulate emotions with artificial facial expressions, vocalizations, and head-nodding movements. Robots thus can identify and construct affective verbal and nonverbal cues conveying approval, denial, attention, comfort, and neutrality, among other microexpressions. As Cynthia Breazeal (2002), a robot scientist, has observed, robots are so verisimilar in their automated facial expressions that human subjects are easily fooled into believing that the automaton is actually communicating with them. In a relevant study, Ciardo et al. (2022) found that by introducing variability in reaction time to the programmed behavior of robots, humans are easily duped into believing that the robot is alive in a human sense. Simply put, the robots were able to perform tasks alongside human partners so convincingly that the latter could not tell whether they were actually human or not.

Epilogue

Prior to its removal from the Internet by Universal Music Group, "Heart on My Sleeve" had become one of the most viewed songs online. As in other areas of AI-generated art forms, the event emphasized that there is no turning back the clock, which ironically is what the Frankfurt School scholars wanted to do in the realm of music already in the 1920s, when popular forms of music starting spreading through the technologies of the era (sound recordings and radio). Music was the central target of criticism of the scholars, because they saw it as the perfect example of how art forms were being commodified to have a short lifespan like mass-produced, assembly-line material products. In contrast, they claimed, true musical art was designed to last, transcending the era in which it was created, because it ennobled the human spirit and could thus not be reproduced rapidly and massively. Despite the many counter-arguments to such cynical views, one nonetheless wonders what the Frankfurt scholars would have to say about AI-generated music and performing robots today.

What the Frankfurt School scholars missed was that some types of popular music can transcend their times, becoming classical in their own way, and thus changing society and aesthetics in the process. As mentioned, jazz initiated a veritable cultural and artistic revolution in America in the 1920s, becoming the musical voice of the Civil Rights movement in the 1950s and 1960s. It is no coincidence that the era of the 1920s was named the Jazz Age. Jazz was rejected at first by the mainstream, denounced as disruptive of American values; but it eventually spread throughout the nation's urban centers, providing the impetus for the emergence of a new and powerful form of culture, open to one and all. A once-condemned music thus came forth to form the aesthetic backbone of American musical art. The seeds of jazz are now everywhere in all kinds of musical forms. One of the most important artists of 1950s rock music, the late Chuck Berry, took the jazz ethos and turned it implicitly into a new loud voice for change. His first hit record, "Maybellene" (1955), established the electric guitar as the new instrument for delivering this voice loudly and excitedly. Berry became an instant musical icon who had a great impact not only on popular music, but also on society at large. His music transcended the era in which it was created.

It is difficult to see how this kind of artistic and social influence can come from AI-generated music, which may indirectly give some substance to Adorno's views that the recycling of materials into a new text is hardly inspirational as music—and AI is clearly a recycling engine of musical forms. The excitement that Berry aroused in listeners can likely never be engendered by machine-made music, no matter how sophisticated the machine is designed to be, at least at the time of writing this book. Referencing Chuck Berry, Aurora Sousanis (2023) makes the following insightful statement, which can be enlisted here to summarize the overall theme of this chapter:

> If you trained an A.I. system on all of Berry's work, it would currently be implausible to expect it to come up with anything other than imitative Chuck Berry songs. Those who fear the future of A.I. should be comforted by the fact that true inspiration is different from data input. Evolution and boundary pushing is (currently) only possible with creative, human, minds at work, rethinking the ways of the past. A.I. is not able to reimagine a world totally unlike any other, but can create in reference to old ideas.

References

Adorno, Theodor W. (1941). *The Culture Industry*. London: Routledge.

Breazeal, Cynthia (2002). *Designing Sociable Robots*. Cambridge: MIT Press.

Chang, Vanessa (2019). Ghost Hands, Player Pianos, and the Hidden History of AI. *Los Angeles Review of Books*, https://lareviewofbooks.org/article/ghost-hands-player-pianos-and-the-hidden-history-of-ai/.

Ciardo, Francesca, De Tommaso, Davide, and Wykowska, Agnieszka (2022). Human-like Behavioral Variability Blurs the Distinction between a Human and a Machine in a Nonverbal Turing Test. *Science Robotics* 7, https://doi.org/10.1126/scirobotics.abo1241

Clarke, Liam (2023). Artificial Intelligence: The Future of the Music Industry? *Recording Arts*, https://recordingarts.com/artificial-intelligence-the-future-of-the-music-industry/.

Coscarella, Joe (2023). An A.I. Hit of Fake 'Drake' and 'The Weeknd' Rattles the Music World. *The New York Times*, https://www.nytimes.com/2023/04/19/arts/music/ai-drake-the-weeknd-fake.html.

Emery, Lynne Fauley (1972). *Black Dance in the United States from 1916 to 1970*. Washington: National Press Books.

Frąckiewicz, Marcin (2023). How AI Is Revolutionizing the World of Classical Music. *Artificial Intelligence, TS2 Space*, https://ts2.space/en/how-ai-is-.

Funk, Tiffany (2018). A Musical Suite Composed by an Electronic Brain: Reexamining the *Illiac Suite* and the Legacy of Lejaren A. Hiller Jr. *Leonardo Music Journal* 28: 19–24.

Gaddis, William (2002). *Agapē Agape*. Harmondsworth: Penguin.

Haraway, Donna (1991). *Simians, Cyborgs, and Women: The Reinvention of Nature*. London: Free Association Books.

Hiller, Lejaren and Isaacson, Leonard (1959). *Experimental Music: Composition with an Electronic Computer*. New York: McGraw-Hill.

Hong, Joo-Wha, Fischer, Katrin, Ha, Yul and Zeng, Yilei (2022). Human, I Wrote a Song for You: An Experiment Testing the Influence of Machines' Attributes on the AI-composed Music Evaluation. *Computers in Human Behavior* 131, https://doi.org/10.1016/j.chb.2022.107239.

Horkheimer, Max (1947). *Eclipse of Reason*. New York: Oxford University Press.

Isacoff, Stuart (2003). *Temperament: How Music Became a Battleground for the Great Minds of Western Civilization*. New York: Knopf.

Lowenthal, Leo (1949). *Triumph of Mass Idols*. New Brunswick: Transaction Books.

Marcuse, Herbert (1972). *An Essay on Liberation*. Harmondsworth: Penguin.

McLuhan, Marshall (1964). *Understanding Media: The Extensions of Man*. New York: McGraw Hill.

Petrarca, Emilia (2018). Body Con Job. *The Cut*, https://www.thecut.com/2018/05/lil-miquela-digital-avatar-instagram-influencer.html.

Reiff, Nathan (2023). Human vs. AI-Generated Music: Indistinguishable, or Uncanny Valley? *Decrypt*, https://decrypt.co/resources/human-vs-ai-generated-music-indistinguishable-or-uncanny-valley.

Schomer, Audrey (2023). Avatars as Actors: Will AI Unleash Celebrity 'Simulation Rights'? *Variety*, https://www.variety.com/2023/digital/news/avatars-as-actors-will-ai-unleash-celeb-simulation-rights-1235583875/.

Solomon, Kate (2023). A Music Critic Reviews AI Songs. *Vice*, https://www.vice.com/en/article/z3mnj4/ai-music-songs-critic-review.

Sousa, John Philip (1906). The Menace of Mechanical Music. *Appleton's Magazine* 8: 278–284.

Sousanis, Aurora (2023). Harmony or Hype? Exploring the Authenticity Debate as Musicians Experiment with AI. *AllMusic*, https://www.allmusic.com/blog/post/harmony-or-hype-authenticity-debate-ai-for-music-creation.

Vogels, Benjamin (2023). Making Music with AI: To Generate or to Compose?—A Music Theorist's Perspective. *Medium*, https://www.medium.com/%40benjamin.vogels/making-music-with-ai-to-generate-or-to-compose-a-music-theorists-perspective-4ce98cf2ad7b.

Wiener, Norbert (1948). *Cybernetics, or Control and Communication in the Animal and the Machine*. Cambridge, MA: MIT Press.

CHAPTER 5

AI in the Mass Media

TV gives everyone an image, but radio gives birth to a million images in a million brains.
—Peggy Noonan

Prologue

In 2023, a popular radio station in Portland, Oregon, called Live 95.5, introduced the first disc jockey powered by AI, named "AI Ashley," whose voice was an AI replica of the voice of the station's human anchor, Ashley Elzinga. After some initial resistance on the part of listeners to accepting AI Ashley, after a while the audience seemed to accept "her" as a co-anchor. The technology used to create AI Ashley is called RadioGPT. It was developed by Futuri Media, located in DeVry University in Ohio. It is a computer system that utilizes the advanced capabilities of neural networks to access real-time events—that is, as they are occurring—and then writing radio scripts to communicate them instantly after to listeners. The AI host also sparked contrasting reactions among radio theorists, especially since the AI-generated voice was so realistic that it was virtually impossible to distinguish it from its human counterpart. In the same time frame, Spotify introduced an "AI guide" with the ability to assess the musical tastes of listeners so well that it could then choose what to play for

M. Danesi, *AI-Generated Popular Culture*,
https://doi.org/10.1007/978-3-031-54752-2_5

them, alongside relevant commentary related to the tracks and the artists, as if the listeners were in a direct dialogue with the human-sounding AI.

Events such as these heralded the arrival of AI radio, constituting a spreading trend, not a quirky technological sideshow, used by stations and podcasts not just to promote themselves but to tap into an ever-broadening acceptance, which can be called a "cybernetic sense," that humans and machines are now interactive agents. Live 95.5's AI Ashley allowed for a new type of radio programming that could tap into the ability of neural networks to keep the machine updated about newsworthy events as they happened, conveying them to audiences in real time. As the comments on Live 95.5's website indicated, AI Ashley became gradually accepted by listeners because of its ability to sound like the human anchor's sibling, as well as its ability to highlight localized content that directly addressed audience interests, thus increasing the relevance and appeal of the broadcasts to local audiences. In an unforeseen way, this event recalled, in microcosm, the importance of radio as the first veritable mass medium motivating the rise and spread of early pop culture in the 1920s, given the fact that the radio medium could deliver content more broadly to all kinds of audiences than any other previous technology. AI radio, however, has gone even beyond this, since AI can autonomously present and generate content, not just transmit it broadly. As a consequence, in the Age of AI the radio medium itself is being reimagined conceptually and communicatively, as are all other media.

The history of modern-day cultures is intrinsically intertwined with changes in mass media technologies. The radio gave people (literate and non-literate) access to all kinds of information and aesthetic events, thus reshaping the "typographic brain," as McLuhan called it (1962), accustomed to reading print for delectation and engagement. It thus retrieved a form of orality, based on the human voice. With the spread of TV as a mass medium in the 1950s the brain was rewired to glean meaning and entertainment from the new visual medium. As neuroscientific research has shown, the visual medium has a direct impact on the processing of information by the brain, different from oral media effects (for example Fields 2016). So, the mass media world in which we are reared have a direct impact on the neuropsychology of understanding. And today the Internet is again rewiring our brains, affecting how we access and process information largely in tidbits (Greenfield 2015). In a phrase, the brain is a plastic organ that is shaped by input from the world. So, as McLuhan so perceptively put it, a new medium "shapes and controls the scale and form of human association and action" (McLuhan 1962: 9).

The "typographic brain," as McLuhan called the brain accustomed to reading print, led to cultural innovations, such as the spread of fiction, as discussed (Chapter 2). In the eighteenth century, with the Industrial Revolution, mass media technologies expanded considerably, establishing the basis for modern society to emerge. Among these were photography, telegraphy, and telephony. By the late 1800s, scientific inventions led to the development of even more powerful mass media technologies, including radio and film. Radio broadcasting systems came onto the social scene around 1919, when the General Electric Company formed the Radio Corporation of America. During the late 1940s, TV stations began operating across various nations, starting to replace radio broadcasting in the mainstream. Hollywood film studios supplemented their profits by supplying the emerging TV networks with programming. Cinema and television became mass media partners—remaining so to this day. In the 1990s, the digital age started becoming a reality and existing media starting undergoing digitization (Jenkins 2006). Now, the appearance of AI Ashley, like the performing classical music robots discussed in the previous chapter, indicates somewhat of a paradigm shift in the relationship of mass media with technology, given that AI is a collaborator with humans in the creation and delivery of cultural content, not just a new medium for content to be communicated.

This chapter will look at AI radio, podcasting, and television. The underlying question that will guide the discussion is as follows: Does this spell the beginning of the end for humans in radio and television broadcasting? As in other domains of culture, at present we seem to still desire that humans be involved in a major fashion on radio and television. The reason, as Lucy Ramos (2023) found after conducting a radio survey, is that listeners "tune in to hear a person on the other side, talking to us." As she concludes: "With our favourite radio hosts, we learn about their lives, their backgrounds and their thoughts and tastes. This can create a deeply personal connection." However, as the Live 95.5 event showed, people might become ever more accustomed to listening to robots, as we engage with them and as the robots increasingly appear to engage with us as they display human-like emotions. This is leading to the tendency to perceive robots as communicants, given our propensity to anthropomorphize robotic voices and the "thoughts" that they are conveying (for a relevant study see Schreibelmayr and Mara 2022).

AI Radio

Although Ashley AI was the first widely publicized robot radio announcer, in actual fact, it was back in 2011 that a virtual disc jockey named Denise was first introduced to the world by a community radio station in Texas. Lacking the technological sophistication of AI Ashley, Denise's voice was nevertheless a reasonable facsimile of a human radio announcer's voice. Denise required human assistance to write radio scripts, but she had the capability to tell jokes on the spot when asked, as well as to provide the weather forecast and look up information on the Internet that was communicated to listeners immediately. Denise lasted only a couple of weeks, being perceived as a kind of sideshow robot performer. However, she started a trend. In 2018, China's state-run news agency, Xinhua, introduced an English-speaking AI anchor whose voice emulated that of a Chinese journalist named Zhang Zhao. It would present itself to listeners as "I am AI news anchor in Beijing." The AI anchor learned from live broadcasting videos by itself via neural network training and had the ability to read texts like a professional news anchor.

With RadioGPT, which produced AI Ashley, the advent of AI radio announcers had finally become a kind of mainstream broadcasting reality. The AI could identify trends by scanning social media platforms and other sources of information; it could write scripts using transformer technology; it could read the scripts on-air with three different voices representing different "radio hosts." In a follow-up survey conducted by Futuri of listeners to the AI anchor, most of the interviewees indicated that they tuned into a radio station to hear a real person on the other side, as if the person was talking to them, and thus were somewhat wary of a synthetic voice on that other side. It also became clear that a desirable feature of human radio hosts was that they tended to discuss their lives, backgrounds, thoughts, and tastes from time to time, thus creating a deeply personal connection with listeners. Nonetheless, overall the AI voice was seen as an acceptable complement to the human one. At around the same time, a Swiss radio station, called Couleur 3, decided to broadcast AI-generated radio programming for 13 hours straight on April 27th, including music produced by algorithms, and deepfake voices of five actual human hosts cloned by AI tools (Diaz 2023). Listeners were reminded every 20 minutes, by one of the cloned voices, that the programming was AI-generated. The station then followed-up with a forum through WhatsApp for feedback from the listeners. The reactions were divided between those who were impressed and those who were appalled by the programming. The head of Couleur 3, Antoine Multone, issued the following statement concerning the experiment afterward (in Diaz 2023):

> It's not easy distinguishing between a human and a robot speaking. However, this experiment also shows us that creativity, surprise, and humor remain human characteristics—and this reassures us, even though we had no doubt about it from the beginning of the project. This day shows us the way forward: working on added value and originality in our content.

It is intriguing to note that this type of comment emerged as well during the early era of radio broadcasting, when people were at first also divided, with some decrying the advent of such technology as the death knell of literacy and even society as it was conceived in the 1920s (Woodford 1929). However, in short time, the radio evolved into the most popular medium for entertainment and information, as well as a force for social unification. As Tom Lewis (1992: 26) has aptly put it: "The first mass medium, radio made America into a land of listeners, joining every age and class into a common culture." Radio could reach many more people than print, not only because it could span great distances instantly through the airwaves, but also because its audiences did not necessarily have to be print literate to engage in the content. Programming was thus increasingly designed with mass appeal. Evidence of a plan for radio broadcasting to the general public can be found in a 1916 memorandum written by David Sarnoff, an employee of American Marconi, which would eventually become the Radio Corporation of America (RCA). In it, Sarnoff recommended that radio be made into a household "utility." His memo was not given any serious consideration by management at first. After World War I ended in 1918, however, several manufacturing companies began to explore Sarnoff's idea for the mass-marketing of home radio receivers very seriously. As a consequence, radios started entering households across the nation.

Radio broadcasting changed social life wherever it was introduced. It brought news, information, and the arts directly into homes. Historically a privilege of the élite, the arts could be enjoyed via the radio by members of the general public, most of whom would otherwise not have the financial ability to access venues such as the concert hall and the theater. The radio thus helped engender an unprecedented mass culture for people of all social classes and educational backgrounds. The "cultural democratization process" started by the Gutenberg revolution in the domain of print technologies, and by the gramophone in the domain of music, was extended by the radio medium considerably.

Now, are we witnessing a second radio revolution, with events such as the AI Ashley one? While early radio united audiences communally, given that only a few stations were available, AI is fragmenting that world, since it can respond to personalized requests, as the Spotify system discussed above indicates. This might get people to listen to made-to-order programming more and more. Such fragmentation would make it almost impossible to establish new cultural trends and paradigms, a fact that may threaten true innovation in pop culture. In effect, AI-generated made-to-order content can potentially impede the advent of new broadly-based movements in the arts—a feature of historical pop culture. As Andy Bhattacharyya (2023) comments on AI-generated music, the clock can no longer be turned back nostalgically; in fact, the "discussion surrounding the impact of AI on music is no longer a novel concept, as artificial intelligence has been making significant contributions to the industry for several years now."

Despite some early criticism of AI radio voices, such as AI Ashley, and the reluctance of some to adopt AI for programming assistance or creation, the radio medium has entered the AI Age in full force. As a result, it will bring with it a host of issues and problems, from ethical to aesthetic ones. More to the point of the present discussion, it may affect how pop culture itself will evolve (or not). One cannot underestimate the importance of the traditional radio to the spread of such culture. Before radio, jazz bands, for example, played the same pieces over many years without becoming old, but as radio broadcasts reached wide audiences, new arrangements, and songs had to be produced at a more rapid pace to keep up with changing tastes and the desire for new material. Radio brought about the commodity culture that the Frankfurt School scholars described (Chapter 4), but at the same time gave music forms like jazz international exposure, allowing the idiom to be recognized as a veritable art form and evolve into an influential code of music-making, as discussed (Chapter 2). Will AI replace human radio announcers? Can AI tell if a new song is truly worthy of being broadcast, or can it only respond to personalized requests? Will anyone who will be raised in a world of spreading AI care? The indifference toward AI announcers is a sign that the ensconcement of a technopoly has been well underway, whereby technology is seen as “autonomous, self-determinative independently of human action, and undirected in its growth, eliminating all other thought-worlds” (Postman 1992: 55). The computer in this world is the “quintessential, incomparable, near-perfect” device establishing “sovereignty over all areas of human experience based on the claim that it thinks better than we can” (Postman 1992: 111).

AI Podcasts

AI technologies are found throughout streaming media. It is in the world of podcasting, where AI has a definite presence. As Kate Knibbs (2023) insightfully characterizes them, AI-based podcasts are a “strange new genre,” which are boosted by the new technologies. One of the first such podcasts to gain popularity was *The Joe Rogan AI Experience*, a series emulating the programs of the controversial podcaster Joe Rogan. The fictional episode was released on YouTube in April of 2023. It consisted of a conversation between deepfake Joe Rogan and Sam Altman, the CEO of OpenAI, with a disclaimer that the “ideas and opinions expressed in the podcast are not reflective of their thoughts.”

While most human podcasters have embraced AI tools, to simplify tasks such as editing and removing background noise, wholly AI-generated podcasts, such as *The Joe Rogan AI Experience*, evoke the same range of questions as do AI radio content and any other AI-generated media content. Podcasting itself is inherently a niche form of media, targeting specific kinds of audiences. AI is used normally to help human podcasters facilitate the creation and delivery of content; but it is in the area of culture-focused AI podcasting the the implications for the future of broadcasting are most consequential. Interestingly, a large number of podcast genres have retrieved many of the functions of early radio, including fiction genres, such as radio dramas and series. But the enormous number of podcasts leads to a splintering of audiences, unlike the old radio shows, which were broadcast on a few networks. According to the Podcast Index database, the number of podcasts is in the millions. As Knibbs (2023) also found in a survey of AI-generated podcasts, the programs are largely indistinguishable from human-made ones, but they do not compete in any significant way with the human podcasts, because it is the personality of the human hosts that make the podcast attractive. As Knibbs (2023) concludes: “the core appeal of tuning in week after week is hearing what the specific humans at the microphone have to say.” In her book *Podcasting as an Intimate Medium* (2023), Alyn Euritt remarks that the emotional power of podcasts inheres, in fact, in how listeners see themselves as "members of an imagined national community," suggesting that the podcast medium is taking on some of the functions of early radio.

However, As in other domains of AI-based media, there are signs that AI podcasting will continue to gain popularity, as audiences become more and more accustomed to android hosts. One such example is Spotify's AI Radio, which is hosted by two chatbots named Bella and Adam, who discuss current events in a simulated humorous way (Xiang 2023). Countless AI podcasts started proliferating at the time of writing this book. And in fact podcasting became early on a kind of experimental laboratory for testing out AI tools in broadcasting and the reactions of audiences to them. Among the reasons why AI podcasting is spreading is the fact that an entire podcast episode can be created in a matter of minutes, compared to traditional podcasting which takes considerable time, and the fact that anyone can create a podcast with no need to have a sophisticated technical know-how.

As AI-generated podcasts continue to gain traction, the main implication to be gleaned is that they offer a glimpse, in microcosm, into the future of media content—how it is produced and how it is consumed, as personalized tastes become ever more a factor in this area, replacing the collectivized tastes of mass audiences of the past, which listened as communities to radio and then television broadcasts. With AI technology, podcasts are able to deliver customized content in a fast and efficient manner, a fact that may again give credence to the notion of the commodification of culture, which formed the basis of the Frankfurt School view of modernity (Chapter 4). It is the massiveness of the cultural artifacts that AI permits, most of which are replicas or combinatorial reproductions of previous human-made artifacts, that leads to a reassessment of the culture industry theory of the Frankfurt School scholars. Proponents of AI-generated culture tout the technical proficiency of programming based on AI, but engagement with the AI content never seems to move beyond that. As Dylan Kull (2023) has perceptively remarked: "AI-generated content is the model product of the modern culture industry. It is mass-producible, generated in seconds without the need for labor, and lacks the necessary DNA of human art, which would preclude its consumption from being in service of or propagating the industry itself." It remains to be seen if the human need for meaningful new content will be subdued by the machine or if it will transcend it.

AI Television

The streaming platform Twitch TV released a sitcom titled *Nothing, Forever* in 2022. It was completely created by AI, and it gained thousands of viewers immediately after it went on-air. The show mimicked the content, humor style, and situational format of the 1990s sitcom *Seinfeld*, in which a group of friends discuss the goings-on of their daily meaningless lives. The original sitcom was in fact dubbed "the show about nothing," and the AI-generated version was designed clearly to pay homage to it, emphasizing that the "nothing" will last "forever." The main character, Larry, is a comedian cloned digitally to resemble Jerry Seinfeld, the original human protagonist of the series. Like the real Seinfeld, the deepfake one does occasional stand-up shows between scenes. There are also "camera shots" of the computer-generated New York City brownstone where he lives, which closely resembles the building of the actual sitcom. The other main characters are Fred, Yvonne, and Kackler, respectively, resembling the original characters George Costanza, Elaine Benes, and Cosmo Kramer. After the show went on a brief hiatus, it returned with a second season with a new cast of characters and a format different from that of the original, but to diminished success, since viewers wanted to see the original sitcom go on forever, as the first season promised. According to comments left on social media platforms, audiences saw the departure from the Seinfeld-based format as absurd, obliterating their nostalgic motivation for viewing the AI-generated redux in the first place.

Perhaps because of the unexpected success of the first season of *Nothing, Forever*, other AI-generated television programs quickly followed, including *AI SpongeBob*, based on the original *SpongeBob SquarePants* television series, which began livestreaming on YouTube and Twitch in 2023. The program used AI-generated models of the characters from the cartoon series, along with cloned voices similar to those of the original voice actors. But, given the technology of the period, the narratives and voices came across as glitched, making the plots fall apart often as incomprehensible. Other negative aspects of the AI series were its profanity and its frequent use of insults against minorities. After being banned a few times, the show kept on appearing on different platforms, but with diminishing viewership.

Another AI program that gained popularity in the same year of 2023 was the first news channel generated entirely by AI, called *NewsGPT*.

There were no human reporters in the program, and the stated aim by the producers of the program was to provide unbiased and fact-based news, which *NewsGPT* was purported to be able to do because it could objectively scan relevant news and information sources from across the world in real-time, using them to create unbiased news stories and reports, by algorithmically extracting a purportedly common ground of meaning in them. But even a small sampling of its newscasts reveals questionable and even inaccurate content—a victim of AI hallucinations? At the time of writing this book, *NewsGPT* produced conspicuously incorrect images of places in the news and seemingly invented statistics to fit the AI's presentation of a news item. So, unlike the claims made by the producers, the AI was actually biased in its own strange, machine-like way, highlighting the problem of so-called algorithmic bias, which results because the algorithm privileges one category over another during its learning phases, largely on account of unintended or unanticipated decisions relating to the way data is coded, selected, or used to train the algorithm. Algorithmic bias has been cited as a factor in cases ranging from skewing election outcomes to the spread of online hate speech. It is beyond the scope of the present work to deal with this vital issue; suffice it to say that it is now being discussed throughout the academic, political, and social spheres. UNESCO put out the following relevant statement (Ramos 2020):

> In no other field is the ethical compass more relevant than in artificial intelligence. These general-purpose technologies are re-shaping the way we work, interact, and live. The world is set to change at a pace not seen since the deployment of the printing press six centuries ago. AI technology brings major benefits in many areas, but without the ethical guardrails, it risks reproducing real world biases and discrimination, fueling divisions and threatening fundamental human rights and freedoms.

Another obvious issue related to AI content concerns copyright and authenticity, given that AI-generated programs, such as those described here, are able to create sophisticated, genre-authentic, fluid scripts that faithfully mimic the original programs such as *Seinfeld* and *Sponge Bob*. As Kilkenny (2023) notes, the question of copyright is truly problematic, given that "copyrighted art does not fall under the fair use defense, which allows for use of protected works without permission as long as they are transformative," especially since the AI may even "bypass the work of a human writer."

Mediated Social Codes

Since the radio medium came onto the scene it has had an indelible impact on society, as mentioned, shaping worldviews, influencing lifestyle, and even the functioning of the mechanisms of perception and cognition involved in distinguishing between reality and its representations. The contents and formats of the radio programs coalesced into an overall system of culture-focused representation of the world, which was encapsulated in their scripts—a system that instituted what can be called a "mediated social code." This code became transferred osmotically to audiences, informing people what aspects of the representations were critical socially and psychologically. In effect, the code came forward to model and emphasize certain ways of thinking and doing things, Social codes have always existed. They emerge to provide guidance on how to live and interact with others. Now, mediated social codes, as the term suggests, come from the media, often being in contrast with existing ones. In a phrase, social codes are grids of communal meaning that assign interpretants to human actions, beliefs, objects, rituals, and so on.

Even a seemingly simple protocol such as greeting someone is coded socially. It is shaped by appropriate words, phrases, and nonverbal cues that will enable a speaker to make successful social contact with another speaker. Consider, for the sake of argument, a simple greeting protocol such as "Hi!" As such, it reveals the kind of coded behavior that is relevant in the society in which it occurs, since there are other societies in which silence or a simple nod of the head may be more appropriate as a protocol. Also, the abbreviated word rather than the full word "Hello!" implies a social relation of a specific kind (informal) between the interlocutors, implying as well that they are involved in similar life schemes, which impel them to make contact in this ritualistic way. Unconscious factors such as these are imprinted in that one simple protocol, thus connecting it to larger frames of meaning and interaction. These frames are coded socially—they constitute ways of relating to others and to oneself. The information exchanged in that one protocol, therefore, goes well beyond a greeting. As this example shows in microcosm, social codes derive from traditions and are imparted *in loco* via rearing practices.

Now, when radio came onto the scene, its programming and formatting system eventually became recognizable in terms of its content, style of delivery, themes, and so on. At first it mirrored the existing moral and ethical codes that society harbored at the time, but through its own self-styled representations it gradually emerged as a shaper of those very codes, unwittingly, of course, even though not everyone read the same social meanings in the programs, leading to debates and conflicts, which came to constitute a "cultural war." Moreover, each new era subsequently produced its own mediated social codes. The way people interacted, dressed, behaved, and even came to view the world were increasingly shaped by mediated social codes—that is, by the representations on radio and then television that produced ever-changing models of how to talk, dress, interact, and even think. The mass media thus evolved into an overarching social code from which modern cultures have extracted their models and meaning-making structures for daily life routines.

The above discussion implies, essentially, that the mass media have been critical agents of social change, for better or worse. Indeed, because television showcased racial protests, riots, and other significant social events, it forced the hand of change several times. Without it, there probably would have been no social justice legislation, and no "accountability" politics. Moreover, many programs and events have been pivotal in bringing about a change in social mindset vis-à-vis certain ethical issues as well as introducing cultural trends that have become part of modern history.

In the current AI Age, the effects of the traditional mass media on the formation or recalibration of social codes have started to weaken because of the fragmentation of audiences (as discussed). So, for instance, a reconstructed sitcom such as *Nothing, Forever*, did not reproduce the socially-coded meanings of the original Seinfeld sitcom, so as to allow them to evolve further, but rather simulated the sitcom out of its temporal context, whereby it lost its socially-coded representational power. The overall meaning embedded in the original Seinfeld code initially reflected a change in worldview and lifestyle in the demographic generation of the 1980s and 1990s, which started to envision the traditional family structure and the courtship rituals of previous generations as lacking relevance to them. This meaning came through in: (1) the dialogues about trivial things; (2) the general lack of interest in marriage and family structure exemplified by the characters; (3) the random events that were spotlighted as part of everyday life, reiterating and reinforcing the program's underlying nihilistic meaning; and (4) the lack of any resolution to the events, which remained trivial throughout the duration of the program. The sitcom had an impact on people's perceptions of social values, embedding itself into groupthink through its episodes, as could be seen in the many discussions and debates it engendered across society at the time. The Seinfeld social code became, over time, a means whereby people in the era could "make sense" of and discuss their own lives, mapping them against the lives of the sitcoms characters.

On the contrary, the AI redux fell flat eventually, because it could not revive the coded meanings for the simple reason that the deepfake sitcom was decontextualized; that is, it did not fit semiotically into the cognitive system of the era in which it was devised (Saussure 1916). The original sitcom was so emotionally powerful for viewers at the time that it continued to have fan clubs and remained popular through reruns, perhaps in a nostalgic way. Now, the humor in the AI version is no longer funny, because it is decontextualized, and the whole program produces a kind of surreal effect on viewers, who are now inclined to read it with different cultural eyes. What was happening inside the original Seinfeld sitcom was happening across the nation, where debates about traditional values were being waged on a daily basis. The Seinfeld social code was meaningful to viewers at the time because it reflected the changes in values and lifestyles that were ongoing in society at the time. Portraying with humor the cultural polarization that had emerged broadly in America. It also highlighted the power of the mass media to both reflect and shape social values.

With AI television the social power of the mass media has become highly diminished, as viewers can get chatbots to create programs in a made-to-order fashion. This has deep implications for the future of pop culture, which has always needed fairly stable and uniform audiences. As Postman (1992) had predicted, as society embraces technology more and more, for entertainment reasons, the conditions for a "socially-code-less" world are ever more real, whereby there are no common values and ideas, but fragmented ones, with people entering into conflict with each other over the content they engage with on niche media—cultural and political.

Each type of technology, and each successive stage, shows an improvement in the efficiency and reach of mass-mediated communications—that is, in the speed, accuracy, and span of such communications. The flow of this technological history reached its zenith with the invention and spread of computer technology—it is this technology that has made the automated and increasingly culturally-fragmented world a reality. It is to be noted that the two problems that communications technologies have tried to solve are the problem of long-distance communication and the transmission of information across subsequent generations. As McLuhan often pointed out, writing provided solutions to both these problems. The written word can reach anyone, anywhere on earth via attendant mass communications technologies. Also, writing is a "time-traveling" artifact. To this day, we can read the words of, say, Aristotle or Confucius, allowing us to go back in time to when they were written thought the imagination. While the interpretation of the words will change, the essence of the ideas they enfold remains fairly stable. In effect, the written word is the foundation of history, encoding it as historiography (or writing about history). Without it, we would lose our connection to the past—each word is a kind of historical time capsule that allows us to re-think the thoughts it originally encoded, in our own retrospective ways. When radio and television emerged as major communications technologies, they shattered the hegemony of print literacy as the unconscious form of recording meaning; AI has shattered that hegemony even further.

Epilogue

Pop culture spread beyond the print medium, with its dime novels, pulp fiction, etc., with the advent of the first mass electronic medium, the radio. Along with cinema and sound recordings, the radio became a primary source of entertainment, emphasizing the role of newer technologies in delivering, embedding, and shaping trends in modern societies, and hence an early congener of mediated social codes. As discussed, mass printing technologies made it possible for the dime novels and pulp fiction magazines that emerged in the nineteenth century to spread broadly; sound recordings made it lucrative to encourage the creation and spread of new musical styles, projecting them onto national and international venues; radio even developed its own derivative genres, such as the soap opera and the sitcom; and television later entrenched the same kinds of genres, even adapting fictional comic book heroes such as Superman, Batman, and others to the new screen.

The argument that the mass media have been critical to modernity is supported by the fact that they have brought about, or at least influenced, changes in politics and society generally. The horrific television images of the Vietnam War broadcast daily in the late 1960s and early 1970s galvanized social protest; the constant treatment of sexual themes on television sitcoms greatly obliterated the silence that previously existed with regard to sexual matters in society at large; the lack of representation of minorities on the television screen eventually led to a social reaction, leading to a more equitable form of representation in the current era. In sum, social mores, politics, and pop culture came to form an intrinsic synergy with the mass media. As Marshall McLuhan (1951: 23) argued, the main aspect of this synergy is that it redirected the creative instinct toward social purposes, away from the idyllic Romantic view of art as self-sustaining and above trivial concerns: "As the unity of the modern world becomes increasingly a technological rather than a social affair, the techniques of the arts provide the most valuable means of insight into the real direction of our own collective purpose."

The question becomes if AI-generated programs and avatar announcers will have completely stemmed the force of the synergy, given that all it takes to create a program is a simple prompt to a computer. One of McLuhan's most relevant insights in this domain is that the "m*e*ssages" of the mass media are actually psychological "m*a*ssages." It is reported that his famous phrase, "the medium is the message," was actually a typographical error in the first edition of his book, *Understanding Media* (1964), which was corrected at some point in the production of his book. His original plan was, in fact, to use the word "m*a*ssage," rather than "m*e*ssage," which conveys more pointedly his theory of media affecting content and social evolution more accurately, as if human minds were being given a massage upon exposure to the mass media. Subsequently, McLuhan used various puns on that word, including "mass age." In his 1967 book, *The Medium is the Massage: An Inventory of Effects*, co-authored with Quentin Fiore, the essence of this view of media is articulated as follows (McLuhan and Fiore 1967: 26), repeated here for convenience:

> All media work us over completely. They are so persuasive in their personal, political, economic, aesthetic, psychological, moral, ethical, and social consequences that they leave no part of us untouched, unaffected, unaltered. The medium is the massage. Any understanding of social and cultural change is impossible without a knowledge of the way media work as environments. All media are extensions of some human faculty—psychic or physical.

The birth and spread of mass pop culture in the 1920s was made possible by a partnership between the mass media and the business-technology world, establishing a synergy that has remained largely intact to this day. The marketplace is where forms of culture (from music to films) are literally "sold" to consumers. This guarantees that the popularity of someone or something rises when the marketplace steps in to give it an economic value, thus turning it into a commodity. This does not, however, impugn the quality of the artistic text or trend; it puts a money figure to it. Critiques of this synergy have always existed, from the Frankfurt School to current anti-media views. But this type of critique may be now moot as AI media are embedding themselves into the pop culture bedrock founded from past media, breaking it up into bits and pieces that may no longer cohere into social codes for future development, as chatbots are being increasingly used by entertainment industries to provide customized services, allowing consumers to control their entertainment experiences more and more. Services like Netflix, Hulu, Spotify, and YouTube use AI algorithms to analyze people's viewing history, preferences, and behaviors so as to provide tailor-made entertainment options. The "massager" is now the algorithm itself, rather than the mediated social code.

Perhaps the most relevant aspect of AI in mediated entertainment is its potential to create content that might challenge traditional notions of creativity. Can AI truly generate comedic texts that stimulate humor and laughter, or just reproduce previous forms of comedy such as *Nothing, Forever*, but which have lost their impact because they are decontextualized? If it did, however, it would still take a human to understand the new form, give it a relevant name, and connect it to the historical flow of culture. As Levi Winslow (2023) has commented on *Nothing, Forever*, maybe the AI humor may become humorous gradually after all in the ever-expanding age of algorithms:

> It's weird, sure, but it's also alluring in its unsettling production. Everything in the show is stiff and artificial. The voice "acting" especially is rough around the edges. Still, *Nothing, Forever* has this strange ability to capture my attention as I watch discount Jerry and friends stand around the apartment, often talking about nothing between spates of empty silence.

References

Bhattacharyya, Andy (2023). Ethics in Generative AI for the Radio and Music Industry. *Medium*, https://uxplanet.org/ethics-in-generative-ai-for-the-radio-jockeys-48925fbc77b6.

Diaz, Maria (2023). Would You Listen to AI-run Radio? This Station Tested It Out on Listeners. *ZDNet*, https://www.zdnet.com/article/would-you-listen-to-ai-run-radio-this-station-tested-it-out-on-listeners/.

Euritt, Alyn (2023). *Podcasting as an Intimate Medium*. London: Routledge.

Fields, R. Douglas (2016). Does TV Rot Your Brain? *Scientific American*, https://www.scientificamerican.com/article/does-tv-rot-your-brain/.

Greenfield, Susan (2015). *Mind Change: How Digital Technologies Are Leaving Their Mark on Our Brains*. New York: Random House.

Jenkins, Henry (2006). *Convergence Culture: Where Old and New Media Collide*. Cambridge: MIT Press.

Kilkenny, Katie (2023). Attack of the Chatbots: Screenwriters' Friend or Foe? *The Hollywood Reporter*, https://www.hollywoodreporter.com/business/business-news/chatgpt-hollywood-screenwriters-film-tv-1235296724/.

Knibbs, Kate (2023). Generative AI Podcasts Are Here. Prepare to Be Bored. *Wired*, https://www.wired.com/story/generative-ai-podcasts-boring/.

Kull, Dylan (2023). AI and the Modern Culture Industry. *Medium*, https://medium.com/%40dhkull6/ai-and-the-modern-culture-industry-9df4f2237903.

Lewis, Tom (1992). "A Godlike Presence": The Impact of Radio on the 1920s and 1930s. *OAH Magazine of History* 6: 26–33.

McLuhan, Marshall (1951). *The Mechanical Bride: Folklore of Industrial Man*. New York: Vanguard.

McLuhan, Marshall (1962). *The Gutenberg Galaxy: The Making of Typographic Man*. Toronto: University of Toronto Press.

McLuhan, Marshall (1964). *Understanding Media: The Extensions of Man*. Cambridge: MIT Press.

Postman, Neil (1992). *Technopoly: The Surrender of Culture to Technology*. New York: Alfred A. Knopf.

Ramos, Gabriela (2020). Ethics of Artificial Intelligence. UNESCO, https://www.unesco.org/en/artificial-intelligence/recommendation-ethics.

Ramos, Lucy (2023). RadioGPT: Should You Be Worried About AI Radio Hosts? *Radio.co*, https://radio.co/blog/ai-radio-hosts.

Saussure, Ferdinand de (1916). *Cours de linguistique générale*. Paris: Payot.

Schreibelmayr, Simon and Mayer, Martina (2022). Robot Voices in Daily Life: Vocal Human-Likeness and Application Context as Determinants of User Acceptance. *Frontiers in Psychology* 13, https://doi.org/10.3389/fpsyg.2022.787499

Winslow, Levi (2023). AI-Generated Seinfeld-Like Twitch 'TV Show' Is Peak Absurdity. *Kotaku*, https://kotaku.com/twitch-nothing-forever-ai-generated-seinfeld-dalle-show-1850061075.

Woodford, Jack (1929). Radio: A Blessing or a Curse? *The Forum*, March 1929, pp. 169–171.

Xiang, Chloe (2023). This Podcast Has Hosts, Jokes, and Expert Guests. It's All AI-Generated. *Vice*, https://www.vice.com/en/article/pkgj38/this-podcast-has-hosts-jokes-and-expert-guests-its-all-ai-generated.

CHAPTER 6

AI-Generated Art

The coming extinction of art is prefigured in the increasing impossibility of representing historical events.
—Theodor Adorno

Prologue

In 2018, a portrait, titled *Edmond de Belamy*, sold at the renowned Christie's auction house for over four hundred thousand dollars. It was made by a French art collective called *Obvious-Art*. The image was created by an algorithm that was trained on 15,000 portraits from the online art encyclopedia WikiArt, spanning the fourteenth to the nineteenth century. The auction claimed that the painting represented a new wave of AI art, and thus a new genre in visual art itself. As Dennis Layton (2023) has perceptively commented on the event: "What is interesting is that this news went largely unnoticed," a result that is supportive of one of the main themes of this book, namely that AI-generated culture is seeping into unconscious groupthink, becoming increasingly nothing more than an option alongside human-made culture. In effect, the auction signaled the advent of yet another paradigm shift in the arts engendered by AI.

The event did however spark debates about the role of AI in the art world and in academia. But most people who viewed the portrait online reacted to it as they would to any human-made portrait, even if the

M. Danesi, *AI-Generated Popular Culture*,
https://doi.org/10.1007/978-3-031-54752-2_6

figure in the portrait, Edmond de Belamy, did not exist, nor was there a human story behind the painting, just an algorithm's ability to generate an artwork that resembled previous ones, using machine learning to do so. AI art is the product of a machine, trained on pertinent artistic data. While we may be inclined to either reject it or accept this "new wave" of art, in some ways it recalls the pop art movement that emerged in the middle part of the twentieth century. Both pop and AI art impel us to reflect on the functions of visual art in human life and to reconsider where the interpretant lies—in the machine (whatever that means), in the observer, or in hermeneutic traditions? These constitute key semiotic issues that will be addressed in this chapter. A relevant anecdote sheds some initial light on why these issues are critical. The 2023 winner of the Sony World Photography Awards in the "creative open" category revealed, after winning, that his entry was actually generated by AI, showing that even experienced art experts could be fooled.

As is widely known, Walter Benjamin (1934) had formulated his famous notion of the "aura" to decry the modern world's penchant for the mechanical reproduction of art forms, claiming that this was part of an ever-spreading destruction of the unique aesthetic and social functions of art, leading to the loss of the notion of originality and of the idea of art as having a unique effect on humans that cannot be replicated mechanically. He would likely today condemn AI-produced art as lacking an aura because it is produced and reproduced by a machine. So, the question becomes: If AI artworks are now being appreciated like human-created ones, with little distinction, where is the aura? Does it even make sense to continue to use this concept? This question will be addressed as well in this chapter, especially since all that one has to do to create an artwork is to prompt an AI to generate, say, a painting in the style of a renowned artist. The machine will then produce a startlingly accurate depiction in a matter of moments. As Kaushik Pal (2023) has aptly observed:

> Art has, until very recently, been the sole domain of human beings; but algorithms now have the capacity to create it, too—which represents a fundamental shift in how we view art. Art is often perceived as a reflection of extremely talented people's skills and thoughts—as though it's beyond the capabilities of the people who don't have such gifts. But with AI, anyone can generate a piece of art—for example, an ornate background pattern—with a few clicks of the mouse.

Pop Art

One cannot overestimate the role of pop art to how we view artworks today. The pop art movement surfaced in the late 1940s and early 1950s with its depictions of commercial objects and pop culture celebrities. The movement was initially a reaction against the pretentiousness and intellectual elitism of abstract art, making brand products, comic strips, and other artifacts (or commodities) of modernity the subject matter of art. The idea was to represent the things of everyday life in a consumerist culture, rather than idealized or abstract subjects and forms. The movement was at first critiqued by other artists and academics, but the public loved the new art because they could easily see their daily lives in them. Eventually, the critics came around and saw in the movement a metaphor for art itself.

Perhaps the best-known pop artist was Andy Warhol, who produced paintings and silk-screen prints of objects such as Campbell's soup cans and images of celebrities. James Rosenquist and Tom Wesselmann used advertising art as the basis of their paintings. Pop art gave assurance that art was for mass consumption, not just for the cognoscenti. Using images that reflected the materialism and vulgarity of modern consumerist culture, the pop artists sought to provide a view of reality that was more immediate and relevant than abstract art. Artists like Robert Rauschenberg and Jasper Johns were among the first to close the gap between art and mass culture. Rauschenberg made collages of household objects and Johns of American flags and bull's-eye targets. Pop art also became intertwined with other movements and events in pop culture. The iconic counterculture rock group, Velvet Underground, founded in 1964, was even managed for a brief period of time by Warhol. The band's provocative musical style, nihilistic lyrics, and unconventional performances set the stage for subsequent music genres, ranging from metal to punk. The band did not achieve great commercial success, given the avant-garde nature of its music, but it became widely acclaimed for its innovative style, mirroring events and styles in the pop art world.

The pop artists, fundamentally, aimed to depict the banality, yet strange hidden beauty, of commodities. Many of the early artists actually came out of the world of marketing and advertising. The difference between advertising design and art became a topic of debate, as Barry Hoffman (2002: 6) remarked perceptively:

> In the hierarchy of cultural criticism the lines drawn between art and advertising started out fairly clear. Art is high. Advertising is low. Art is elite and refined. Advertising is vulgar and democratic. Art is original. Advertising is derivative. Art is a product created by people to express their personal vision. Advertising is created by people who get paid to sell a product. Art is defined by the truth of the insight it expresses. Advertising expresses the insight of truisms. Art is disturbingly honest. Advertising is only as honest as it has to be, and occasionally less. Art is eternal. Advertising is ephemeral.

For pop artists, the factory, the supermarket, and garbage can, the comic book, and commercial advertising, became their art school. But despite its apparent crassness, bordering on absurdity, pop art gained acceptance in the art world and was liked by people generally (as stated), perhaps because it gave common people the assurance that modern art was for everyone. The pop artists sought to provide a view of reality that was more immediate and relevant than that of past representational traditions or of the modernist expressionist and abstract artistic movements. They wanted the observer to respond directly to the object, rather than to the skill and viewpoint of the artist. As Hoffman (2002: 107) goes on to observe:

> Pop, like advertising, is interested in the concept more than the rendering. It uses the objects that inhabit the world every individual of every class takes for granted—the mundane, mass-produced stuff that is all around us. The things you use and like. Pop artists don't use these things because there is nothing else to paint, they use them to make a point.

An example of a pop art work that gained wide acclaim was *Just What Is It That Makes Today's Home So Different, So Appealing?* (1956) by the British artist Richard Hamilton. In this satiric collage of two ludicrous figures in a living room, the pop art hallmarks of crudeness and irony are emphasized. As art historian John-Paul Stonard (2007) commented on the collage, the principal template was an image of a modern sitting room in an advertisement in *Ladies' Home Journal*, and its commodified aesthetic. Roy Lichtenstein also became widely respected for his mass-produced commercial illustrations, especially comic strips and satires of advertisements, which drew attention to the most common images in daily life. His comic strip paintings included words simulating portions of dialogues, as the speech bubbles in real comic strips. Pop art developed rapidly during the 1960s, as painters started to focus their attention on brand-name commercial products, producing sculptures of hamburgers and other fast-food items. In the early 1960s Warhol carried the idea a step further by adopting the mass-production technique of silk-screening, turning out hundreds of identical prints of Coca-Cola bottles, Campbell's soup cans, and other familiar commercial products, including identical three-dimensional Brillo boxes.

One fascinating aspect of AI art is its ability to generate pop art-style images from plain text (Fig. 6.1).

The AI-portrait is highly reminiscent of the kind of portrait of Marilyn Monroe that Andy Warhol created and for which he became famous. It is a celebration of the 1950s–1960s pop celebrity "look" and is thus a

Fig. 6.1 AI-Portrait, Emilie Arnoux (Wikimedia Commons: Creative Commons Attribution-Share Alike 4.0 International)

throwback pop art text. Given the fact that AI can generate any imitative pop art portrait, on personal demand, and leaving aside the ethical and copyright ramifications of using generative AI in this way, the relation of such art to human originality and creativity stands out as the most relevant aesthetic, psychological, and social issue. One especially concerning area is that of deepfakes. For example, in 2018, BuzzFeedVideo created a convincing deepfake of President Barack Obama, that also mimicked his voice and gestures. This quickly raised ethical concerns among viewers, even though the actual message was to become wary of the disinformation deepfakes could create. But then, in a real sense, Warhol's screen paintings of Marilyn Monroe and other celebrities were deepfakes themselves, created by a human rather than an algorithm. Art-generating machines can easily create portraits and objects that look like the people and things in the world, and it is thus hard to see why this is not art in itself.

AI Art

The *Edmond de Bellamy* portrait mentioned at the start of this chapter was created with an algorithm called Generative Adversarial Network, which has the ability to imitate painting styles from any era. Below is the portrait, which can be seen to imitate impressionist painting style, as suggested by the blurriness of the image. Note that the signature in the bottom right corner of the canvas is, rather appropriately, the mathematical formula of the function that was utilized to produce it: $\min G \max Dx\left[\log(D(x))\right] + z\left[\log(1 - D(G(z)))\right]$ (Fig. 6.2).

Assigning meaning and aesthetic relevance to this portrait is not a simple matter of utilizing the traditional canons of hermeneutic analysis, which imply assessing the relevance of the era in which the work was created, the particular stylistic characteristics of its painter, and its relevance to the world of art itself. Now, *Edmond de Bellamy* did not emerge in a historical era of art; the painter was an algorithm; and it is hard to locate its hermeneutic relevance to art beyond the fact that it appears for all intents and purposes to be a legitimate and "meaning-bearing" portrait. All that one can realistically say about such a work is that ultimately it is the audience that decides whether to accept it as art or not, no matter who or what created it. Since it garnered a lot of interest at the Christie auction, it can be said to have struck an aesthetic chord, much like any human-created portrait. This implies that the aesthetic notion

Fig. 6.2 AI-Portrait, *Edmond de Bellamy* (Wikimedia Commons, https://commons.wikimedia.org/wiki/File:Dmond_de_Belamy.png) (Attribution: "This file is in the public domain because it is the work of a computer algorithm or artificial intelligence and does not contain sufficient human authorship to support a copyright claim")

of originality, as distinguished from reproductions, clones, forgeries, or derivative works, requires revision, at the very least.

One never knows what the machine will produce, when prompted. So, in a sense, it is "creative," in terms of the unexpectedness of its production. The makers of the portrait above simply decided what data to use to train the AI model, but it was the computer that ultimately generated the artwork, which was original in its own way. Consider two widely used AI platforms for generating artworks, Midjourney and DALL-E. The former is trained on specific styles; but in the end it comes up with its own creative product, which may even defy classification according to style, genre, or era, as can be seen in its painting titled *Théâtre d'Opéra Spatial* (Fig. 6.3).

The painting won a fine art competition at the 2022 Colorado State Fair, evaluated as possessing true originality by judges, who were unaware

Fig. 6.3 *Théâtre d'Opéra Spatial*, Midjourney, 2022 (Wikimedia Commons, https://commons.wikimedia.org/wiki/File:Théâtre_D'opéra_Spatial.png) (Attribution: "This file is in the public domain because it is the work of a computer algorithm or artificial intelligence and does not contain sufficient human authorship to support a copyright claim")

that Midjourney was used to generate the work. Interestingly, the same judges later commented that even if they had known this, they would have awarded the top prize to the painting anyway. What mattered was the work itself, which shows a scene of women wearing theatrical dresses in an opera setting, but looking out into a portal that opens up into space. It is this juxtaposition of time-divergent events that gives the painting a surreal feeling that blends the past with the future. Clearly, it is the artistic text itself that determines how it will be interpreted, not the author's autobiography, the era in which the art piece was created, or any decipherment of a specific painter's intent—all of which AI art lacks. Once the machine had been trained to generate this painting, mixing and matching various features from previous art, it clearly produced an entirely new piece. This, by the way, is essentially what human artists do during their own training in an art school setting. But in the school, the training is focused on activating the students' *fantasia*, while in AI training the focus is on getting the machine to combine elements in the database to produce a composite, not to activate the algorithm's imagination.

At this point it is useful to adopt Margaret Boden's (1990) division of creativity into two areas: (1) "P-creativity" (or "psychological creativity"), which implies creativity that is novel to the agent that produces it; and (2) "H-creativity" (or "historical creativity"), which implies creativity that is recognized as novel by society at large. AI art can only fall into the category of H-creativity, since it does not recognize its own work as P-creative, at least as far as can be told. Boden also differentiates between the creativity that arises from an exploration within an established artistic code, and the creativity that arises from a deliberate transformation or transcendence of that code. She labels the former as exploratory creativity and the latter as transformational creativity, seeing the latter as a form of creativity far rarer than the former. Boden's insights have guided work in so-called computational creativity at a general level. To paraphrase Boden's overall approach, the machine can be said to be creative in constrained ways, whereas human creativity enfolds all possible forms, with no constraints, given that it can break existing artistic codes and still produce meaningfulness.

As another example of widely a discussed AI artwork, consider the following painting, created by Stable Diffusion (a deep learning model), which was produced by the algorithm from the following simple prompt: "an astronaut riding a horse, by Picasso" (Fig. 6.4).

If one did not know that the painting was generated by a machine, one could easily interpret it as an unknown work by Picasso, with an incongruous twist, given the astronaut figure in it. Stable Diffusion's ability to model the stylistic characteristics of Picasso impels us, evidently, to interpret the work as, perhaps, something that Picasso had not released to the public (hence a posthumous work) or else a human artist paying retrospective homage to Picasso, not as a work by an algorithm. Interestingly, when the machine is asked to interpret its own work, it comes up with weirdly similar explanations to those just outlined, given that it summarizes the type of criticism of Picasso's art from its training database.

AI art clearly raises profound questions about creativity and the nature of art, inducing us to locate any "meaning" that a work might have in the text itself and according to the context in which a work is viewed—hence

Fig. 6.4 Painting made by Stable Diffusion (Creative Commons CC0 1.0 Universal Public Domain Dedication. https://en.wikipedia.org/wiki/File:An_astronaut_riding_a_horse_%28Picasso%29_2022-08-28.png)

the notion of floating interpretant, which is similar to the idea put forth by Barthes in his key work *Image-Music-Text* (1972) whereby "all images are polysemous; they imply, underlying their signifiers, a 'floating chain' of signifieds which allows the reader to choose some and ignore others" (Barthes 1972: 38–39).

Consider one last AI work, *A Painting of a Hand in the Style of Van Gogh*, generated by DALL-E (Deep Artificial Language Learning Engine), which has the ability to edit existing images by adding or removing elements from them, producing some truly remarkable paintings (Fig. 6.5).

However it was produced, this painting can only be described as a "work of art." Perhaps in the same way that AI fiction may be signaling the "Death of the Author," AI art may be similarly signaling the "Death of the Artist" (Deresiewicz 2020). As discussed (Chapter 2), it was Roland Barthes who used the former terminology to indicate that the only realistic interpretation of a work of literature is the one assigned to it

Fig. 6.5 DALL-E's Generation of A *Painting of a Hand in the Style of Van Gogh* (Wikimedia Commons, https://commons.wikimedia.org/wiki/File:A_painting_of_a_hand_in_the_style_of_Van_Gogh_-_DALL-E.png) (Attribution: "This file is in the public domain because it is the work of a computer algorithm or artificial intelligence and does not contain sufficient human authorship to support a copyright claim")

by readers at different periods of time (Barthes 1968). Once a literary text is published, it takes on a life of its own, becoming open to an open set of floating interpretants. The tradition of viewing the author as a singular figure is a convenient western hermeneutic construct, according to Barthes, which is in contrast to other traditions across the world where authorship is not similarly crucial—all that counts is the work itself. This was certainly true in antiquity, when no one knew who the makers of the origin myths were; all that counted was the message in the myths themselves. Similarly, no one knew who the sculptors of statues in the public squares were; again, all that mattered was what they meant or symbolized to people at the time.

While it is true that the intentions of individual artists and the historical context in which they produced their art shaped their creative efforts, ultimately, the value of their works depend on their ability to transcend the time periods in which they were devised. The concept of interpretant, again, is the relevant one here. There are three levels associated with the interpretant (Peirce 1931–1958), already discussed, but which can be reiterated here and adapted to the present discussion as follows: (1) the immediate interpretant which refers, essentially, to the immediate reaction we get from DALL-E's painting; (2) the dynamical interpretant, which is what the viewer gradually comes to grasp about the meaning of the painting; and (3) the normal interpretant, which refers to the meaning that the painting was designed to elicit in conventionalized ways (such as critical assessments). In AI-generated art, the normal interpretant is clearly, in the critical assessments of humans viewing the painting.

It is relevant to note that Barthes (1968) began his essay with a passage from the novel *Sarrasine* by Honoré de Balzac (1819), asking: Who 'speaks' those words: the hero of the novel, or Balzac himself? If it is Balzac, is he speaking personally or on behalf of all humanity? But then why would anyone ask that question in the first place? Any such question does not emerge at the level of the immediate or dynamical interpretant, but mainly at the normal level, which is where it belongs. Barthes goes on to point out that some writers, such as the nineteenth-century French poet Stéphane Mallarmé, reminded readers that it is the *langue* which speaks to them, rather than he himself, the author, recalling Saussure's (1916) notion of a system of signs to which we all have access, if we know the relevant code. So, when reading a novel by Balzac or viewing a painting by Van Gogh we are interpreting the *langue* (verbal or visual) with which the works were created, not the intentions of the creator. In this framework, Midjourney or DALL-E utilize the same kind of *langue* to produce their texts—hence their interpretability in human terms. Interestingly, in the area of graphic design, we rarely apply the same critical standards as we do to "artworks," implying that the requirement of authorship is a conventionalized one, rather than a necessary one (Skaggs 2017)? We tend to interpret design as a functional *langue*, whereas we interpret textual artifacts such as paintings and sculptures as having inherent value in themselves beyond their *langue*.

The Aura

In his seminal and widely-cited essay, "The Work of Art in the Age of Mechanical Reproduction," Walter Benjamin (1934) put forth the idea that a work of true art had a special quality, called an "aura," that cannot be reproduced. It is an unconscious sense of beauty that evokes a feeling of transcendence (Bigliardi 2018). The aura is precisely what cannot be reproduced in a work of art: its transcendent presence beyond time and space, an idea that dovetails with the notion of time-binding by Korzybski (1921), mentioned previously (Chapter 1), at the same time that it transcends temporality—a paradox that has never been resolved by any theory of art. Benjamin suggested that a work of art loses its aura when it becomes difficult to determine when, why, and how a piece of art is created. This means, in terms of AI-art, that mechanical reproduction cannot possibly encompass an aura. As Benjamin (1934: 4) phrased it: the "sphere of [artistic] authenticity is outside the technical [sphere]" of mechanized reproduction.

Applying Benjamin's notion to AI art concretely, it becomes saliently obvious that within this critical framework it is moot to say that a work such as the *Edmond de Bellamy* portrait has, or has not, an aura; the reason may well be, as Will Jennings (2022) states, referencing Benjamin, that "Anybody, even those with little ability in representational painting or drawing, can deploy AI generators to summon fleeting creative ideas, but any user should be aware that the billions of dataset images their images are formed from were all created by makers, artists, illustrators, photographers, and other human creators." As such, AI art is simply a matter of grafting preexisting images to create a new one, via a guided algorithmic process. With combinatorial reproducibility, the aura, as Benjamin envisioned it, would deteriorate. Nonetheless, as discussed, we still get a sense of unique meaning in viewing AI art, implying clearly that the aura is not in the work itself but in the viewer.

So, by just looking at an algorithmically-produced work such as the *Théâtre d'Opéra Spatial*, an aura seems to emanate from it in a unique way. This poses a veritable dilemma as to what art is, especially in the AI Age. Moreover, since Benjamin defines the aura as emanating from a work that enfolds "the here and now" (Benjamin 1934: 15), one can argue that this is exactly what a work of AI art encompasses. In fact, AI does not replicate an art form; it creates one via combinatorics. The aura was located *there,* in the painting, when human judges awarded Midjourney's

Théâtre d'Opéra Spatial. Perhaps it was the context in which the painting was presented—an art competition—that impelled the judges to view the work on its own merits, no matter who or what produced it. This implies that the culturally-based context in which a work is assessed is critical in shaping the perception that it is art or not.

When people are asked *what* any painting hanging on a gallery wall means, no matter how strange it might appear, as is the case with most abstract-expressionist paintings, they tend to find meaning in it because it is *located* in a gallery. Consider a soda can found thrown away on a city street. If we were to come across this item on the street, we would no doubt view it as a piece of garbage or rubbish. But if we saw the very same object on a pedestal, displayed in an art gallery, signed by some artist, and given a title such as "Waste," then we would interpret its meaning in a vastly different way. We would, in fact, be inclined to interpret it as an artistic form, descrying a throw-away or materialistic society. Clearly, the can's location on a sidewalk versus its display on a pedestal in an art gallery, will influence how we will interpret it. Now, does the can have an aura? The fact that such "art" would not evoke the same response all over the world suggests that the aura may not be something intrinsic, but rather a notion acquired in a larger cultural context. The modern idea of visual art as something transcendent started taking shape in the late Renaissance. Only after the Romantic nineteenth century did the idea of art as a unique form of creativity emerge as an *idée fixe*.

Perhaps Vico's notion of the *fantasia* (Chapter 1) can help shed some light on the question of what art is. The *fantasia* is the ability to transform sensory or affective input into figments of inner meaning, which can then be given physical form in creative ways, to greater or lesser degrees. In effect, the true sculptor, for instance, is able to flesh out meaning from formless matter with the *fantasia*. In effect, the human form is imagined by the artist in a slab of marble, who will then literally mold it with the hands. Prefiguring the biologist Jakob von Uexküll (1909), Vico found a constant point of contact between the *fantasia*, as a form of "inner vision" with actual vision (the act of seeing with the eyes). Von Uexküll argued that every species had different inner and outer "visions." In humans the inner visions are unique, since they are produced by the *fantasia*. Now, when someone is able to turn these visions into physical forms such as sculptures or paintings, then our tendency is to recognize them as reflecting our own inner visions, since they evoke our own *fantasia*. The American philosopher Susanne Langer (1948: 129) called this process, appropriately, a "fantasy."

The relevant question becomes: Does AI have a *fantasia*? This is a rhetorical question of course, because AI cannot possibly have the same faculty as humans. It would need a biological body and it would need to have literally "lived" through the body. Moreover, it would need to have a sense of history, as Vico maintained, suggesting that the *fantasia* was the force behind the flow of history—an imagined retelling of the past. This brings us back to the paradox of art. Art does indeed have an "aura"; but then what is it? Using Susanne Langer's (1948) distinction between presentational and discursive understanding, it can be suggested that on a first viewing of an artistic work we get an intuitive sense of its meaning—hence, the aura. Langer called this the presentational form of interpretation. When we attempt to explain our feeling state, however, we are forced to communicate it in terms of language, which Langer called this the discursive form of interpretation, which will never be able to cover the feeling effects produced by a work of art. If those effects are shared broadly, then the work will transcend the period in which it was created, changing everything. The course of human history was changed permanently after Michelangelo sculpted his *David*, Mozart composed his *Requiem*, Beethoven composed his Ninth Symphony, and so on and so forth. The aesthetic effects in such works are constantly being experienced across time. It remains to be seen if an AI-generated work will also be transcendental in a similar way and thus affect the course of future history.

Algorithmic Art Criticism

The foregoing discussion has focused on human assessments of AI art. The question now becomes: Can AI develop its own viable theory of art, which can help resolve the paradox described above? A related question is: Can AI be its own critic and thus produce an algorithmic theory of its own art? Consider the relevant work of Ahmed Elgammal and his team of researchers (see Elgammal 2019), who developed an algorithm that was able to assess the creativity of any given painting, taking into account the painting's historical context. The researchers used computer vision to build a database of paintings from the fifteenth to the twentieth centuries. Through a series of mathematical techniques, the team showed that the problem of creativity could be reformulated as a variant of network centrality problems, which are widely used in the analysis of social interactions and web searches. The input was what the algorithm was trained on via the paintings—namely color, texture, perspective, and subject matter. The parameter setting was the definition of creativity as something original with lasting influence.

The experimenters ran an analysis of 1700 paintings, coming up with several notable findings. For instance, the algorithm scored Edvard Munch's *The Scream* (1893) much higher in creativity than its later counterparts. The algorithm also assigned Picasso's *Ladies of Avignon* (1907) the highest creativity score of all the paintings it analyzed between the years 1904 and 1911. Significantly, these assessments are in line with those of art historians, suggesting that not only can algorithms generate artworks, but also that they may be viable "art critics." However, the researchers did find that the algorithm's assessments did not always coincide with those of art historians. So, they conducted what they called "time machine experiments," in which the date of an artwork was changed to some arbitrary point in the past or in the future, and then got the algorithm to recompute its creativity scores. The findings were truly remarkable. The researchers found that paintings from the Impressionist, Post-Impressionist, Expressionist, and Cubism eras saw a significant increase in their creativity scores, when moved back to around 1600. In contrast, Neoclassical paintings did not gain much when moved back in time to the same era, which is understandable, since Neoclassicism is considered a revival of the Renaissance. Paintings from the Renaissance and Baroque eras experienced losses in their creativity scores when moved forward to 1900.

The results imply two main things: (1) the temporal context may indeed constrain the meaning of a work of art, and (2) humans may not be the only judges of artistic creativity, since algorithms can seemingly perform the same task. The goal of traditional criticism within various humanistic disciplines is to infer what a work of art means, which is done subjectively by a human critic. This often leaves gaps in the interpretation process and, more seriously, may even lead to a nonsensical interpretation of a text, based on some ideological bias or personal ax to grind. AI not only avoids this kind of solipsistic subjectivity, but also unexpectedly penetrates the interpretive subtexts in ways that were never envisioned, perhaps providing insights into the behaviors and inner thoughts that the text-makers were likely to have when they created their texts, as Neuman et al. (2023) have shown with the use of AI in interpreting texts from all areas of human creativity.

The question now becomes: Are there any AI programs designed specifically to be "art critics?" One such program (at the time of writing this book) is StyleGAN, which can generate interpretations of artworks; another one is the Artmatrix algorithm, developed by a team of researchers at Rutgers University. Artmatrix has the ability to analyze an image and provide a creativity score based on several factors, including color, texture, and composition. Now, while algorithms can generate creativity scores, they cannot truly interpret the meaning behind an artwork, which is a product of intersecting contextual factors (historical, psychological, sociological, etc.). Overall, however, as Bahadir Guner (2023) aptly observes: "The development of AI art critics raises questions about the role of the human critic and whether AI analysis can provide a unique perspective."

Semiosis

At the time of writing this book, an AI art-generating system was largely dependent on "humans giving it new content and telling it how to interpret that content, not to mention programming the model and building, maintaining and powering its hardware" (Nelson 2023). Now, if AI were able to program itself from scratch then it would imply that AI art, created without any form of human guidance, would be a truly paradigmatic event that would permanently alter the course of the role of art in human history. It would make AI a new (cybernetic) species, with which we would have to learn how to interact in new ways, perhaps even learning its artificial language, as it has evolved *within* the machine.

This possibility brings us back to the Peircean concept of *semiosis* (Chapter 1). In humans, this implies, above all else, the capacity to transform information into cultural codes and the representations that these allow (such as artworks), which then guide human actions and life schemes, producing what the Estonian semiotician Juri Lotman (1991) called the "semiosphere," a domain of semiotic codes that imbue the human world with structures of meaning (words, symbols, drawings, etc.) and which enhance understanding of the world and even survival. As the late Hungarian-American semiotician Thomas Sebeok (2001) suggested, semiosis is the litmus test to determine whether something is truly alive and sentient, or not. The question becomes: Can an inanimate entity, such as AI, pass what can be called a semiosis test?

Semiosis is an innate trait, manifesting itself early on in infancy. When children come into contact with a new object, their instinctive reaction is to explore it with their senses—that is, to handle it, taste it, smell it, listen to any sounds it makes, and visually observe its features. This exploratory phase constitutes an instinctive sensory state of understanding, which produces memorable images of the object in the brain which then allows infants to recognize the same object subsequently without having, each time, to examine it sensorially over again (although they might explore its physical qualities further for various other reasons). Now, as children grow, they start to engage more and more in semiotic behavior that displaces this instinctive sensory phase; that is, they start pointing to the object and imitating any sounds it makes, rather than just handling it, tasting it, etc. These imitations and indications are the child's first attempts at *modeling* the world semiotically, rather than sensing it instinctively, projecting an object into the domain of inner consciousness. Thereafter, the repertoire of semiotic modeling activities increases dramatically, as children learn in context how to refer to the world through language and the other codes to which they are exposed, rather than through the senses or the instincts alone.

Semiosis reveals, in effect, how the brain carries out its work of transforming sensory information into conceptual knowledge. The specific signs (words, symbols, etc.) that we acquire in childhood subsequently shape how we view the world. This does not mean that understanding between members of different cultures is blocked. On the contrary, as Lotman claimed, we are all part of the same semiosphere, with only localized variations to its semiotic unity. Do the same principles of semiosis apply to AI? Can the AI understand itself? Does it have the ability to understand the meaning of the art forms it creates as saying something about itself and the world? Whatever the truth of the matter, it is obvious that if AI would truly evolve into a self-sustaining species, its representations would require a new language to be understood by us humans, radically transforming the hemisphere.

Epilogue

The pop art movement dovetailed with the spread of pop culture as a mainstream culture by the 1940s. A defining characteristic of pop culture has always been that it reflects changes in the intellectual and artistic Zeitgeist. As mentioned, the Velvet Underground joined forces with Andy Warhol to create an artistic bricolage. But even before this, George Gershwin blended jazz with classical style in works such as *Rhapsody in Blue* (1924) and his marvelous opera *Porgy and Bess* (1935).

Perhaps a work that saliently symbolizes the intrinsic relation between pop art and popular culture is the Beatles' 1967 album, *Sgt. Pepper's Lonely Hearts Club Band*. It was one of the first concept albums of rock and roll. Previously, a rock album would contain a collection of songs by an artist, with no thematic link among them. But *Sgt. Pepper's* changed all that, once and for all. Although concept albums go back to the 1930s, for rock and roll this was a veritable innovation. The cover was itself revolutionary. It was designed by two pop artists, Peter Blake and Jann Haworth, from a sketch by Paul McCartney (one of the members of the band). The band is featured standing behind a drum skin. In front is an arrangement of flowers spelling out the name "Beatles." The group members are dressed in glowing multi-colored military-style uniforms. We can also see wax sculptures of the band members, along with photographs depicting a diversity of famous people, from actors and sports figures to scientists, religious and political leaders, and great writers. This is truly a work in classic pop art collage style.

It could well be that, parallel to how pop art became intertwined with mainstream popular culture, so too AI art may have already become interwoven with ongoing changes in how we perceive cultural representations. The following comment by Aditi Bainss (2023) encapsulates the implications of the crossroads at which we now find ourselves:

> The intersection of AI and pop culture is a fascinating area of study that highlights how technology is changing how humans interact with media and each other. As AI becomes more sophisticated, it will continue to shape and influence pop culture in new and exciting ways. From AI-generated music and art to hyper-realistic fashion imagery, we are already seeing the impact of AI on pop culture. As we progress, we must consider the potential positive and negative effects of these changes on human behavior and attitudes. By embracing the potential of AI in pop culture while being mindful of its limitations and drawbacks, we can work towards

a future where technology and culture are more inclusive, representative, and meaningful for all.

This is an optimistic view, which is not shared universally. Nonetheless, it is true that, as the pop art movement signaled a shift in how art was being viewed in a pop culture environment, so too it can be asserted that AI art is signaling a similar kind of shift, which may indeed be signaling how pop culture is changing "in new and exciting ways." In fact, with AI we may no longer need human pop artists. As Layton (2023) observes, the advent of AI art has clear implications that can no longer be ignored: "Creating art that no one has seen before, influenced by the works of others, is no longer a uniquely human ability." On the other hand, as Layton states right after: "creating art that in some way or another transcends what has been seen before likely will always be the work of a human artist."

References

Bainss, Aditi (2023). Exploring the Intersection of AI and Pop Culture. *Rephrase.ai*, https://www.rephrase.ai/blog/exploring-the-intersection-of-ai-and-pop-culture#:~:text.

Balzac, Honoré de (1819). *Sarrasine*. Paris: Charles Gosselin.

Barthes, Roland (1968). Death of the Author. *Aspen*, issues 5–6, https://www.ubu.com/aspen/aspen5and6/threeEssays.html.s.

Barthes, Roland (1972). *Image-Music-Text*. New York: Fontana.

Benjamin, Walter (1934). *Gesammelte Shriften*: Frankfurt: Suhrkamp Verlag.

Bigliardi, Victoria (2018). The Reincarnation of the Aura: Challenging Originality with Authenticity in Plaster Casts of Lost Sculptures. In: Marcel Danesi (ed.), *Empirical Studies on Semiotics and Visual Rhetoric*, 299–318. Hershey, PA: IGI Global.

Boden, Margaret (1990). *The Creative Mind: Myths and Mechanisms*. London: Weidenfeld & Nicolson.

Deresiewicz, William (2020). *The Death of the Artist*. London: Macmillan.

Elgammal, Ahmed (2019). AI Is Blurring the Definition of Artist. *American Scientist* 107(1): 18–22, https://doi.org/10.1511/2019.107.1.18.

Guner, Bahadir (2023). The AI Art Critic: What Happens When a Robot Reviews Art. *Medium*, https://www.bahadirguner14.medium.com/the-ai-art-critic-what-happens-when-a-robot-reviews-art.

Hoffman, Barry (2002). *The Fine Art of Advertising*. New York: Stewart, Tabori & Chang.

Jennings, Will (2022). The Work of Art in the Age of Artificial Production. *Tribune*, https://tribunemag.co.uk/2022/11/the-work-of-art-in-the-age-of-artificial-production.

Korzybski, Alfred (1921). *Manhood of Humanity: The Science and Art of Human Engineering*. New York: E. P. Dutton & Company.

Langer, Susanne (1948). *Philosophy in a New Key*. Cambridge: Harvard University Press.

Layton, Dennis (2023). AI Art and the Portrait of Edmond Bellamy. *Medium*, https://medium.com/%40dlaytonj2/ai-art-and-the-portrait-of-edmond-de-belamy-43f318882c09.

Lotman, Juri (1991). *Universe of the Mind: A Semiotic Theory of Culture*. Bloomington: Indiana University Press.

Nelson, John P. (2023). ChatGPT and Other Language AIs Are Nothing Without Humans. *Scientific American*, https://www.scientificamerican.com/article/chatgpt-and-other-language-ais-are-nothing-without-humans.

Neuman, Yair, Danesi, Marcel, and Vilenchik, Dan (2023). *Using AI for Dialoguing with Texts: From Psychology to Cinema and Literature*. London: Routledge.

Pal, Kaushik (2023). 5 Ways AI is Changing Art. *Techopedia*, https://www.techopedia.com/what-is-the-impact-of-ai-on-art/2/33399.

Peirce, Charles Sanders (1931–1958). *Collected Writings*, 8 Vols. Cambridge: Harvard University Press.

Saussure, Ferdinand de (1916). *Course de linguistique générale*. Paris: Payot.

Sebeok, Thomas A. (2001). *Signs*. Toronto: University of Toronto Press.

Skaggs, Steven (2017). *Fire Signs: A Semiotic Theory of Graphic Design*. Cambridge: MIT Press.

Stonard, John-Paul (2007). Pop in the Age of Boom: Richard Hamilton's 'Just What Is It That Makes Today's Homes so Different, so Appealing? *The Burlington Magazine* 149: 607–620.

Uexküll, Jakob von (1909). *Umwelt und Innenwelt der Tierre*. Berlin: Springer.

CHAPTER 7

AI in Marketing and Advertising

AI is probably going to be the single biggest disruptor and enabler of marketing in total. It's a dream come true, from a marketer's perspective.
—Raja Rajamannar

Prologue

The company called MSCHF (pronounced "mischief") introduced an experimental fashion brand of boots in February of 2023. In a press release on its website (https://mschfshop.com/product/big-red-boots/), the following statement was published: "Cartoonishness is an abstraction that frees us from the constraints of reality. If you kick someone in these boots, they go boing!" MSCHF is an American art collective based in Brooklyn, New York. It has produced a wide range of brands, artworks, and platforms, from sneakers and foot photographs to social media channels—all based on AI. The boots became an instant fashion craze, despite their truly cumbersome and gross cartoonish appearance—a fact that clearly bears implications about how consumerist culture, fashion, and even pop culture are evolving in the AI Age. As Christian Allaire (2023) remarked on the craze:

M. Danesi, *AI-Generated Popular Culture*,
https://doi.org/10.1007/978-3-031-54752-2_7

> Given how impractical and unusual the footwear is, it's been a surprise to see how quickly—and how widespread—the fashion world has hopped on them, before they're even released no less. They're certainly a commitment to wear: A number of TikTok videos show how much of a workout it is to get them on and off. But if fashion people love one thing, it's a gimmick, and these boots had everyone talking and noticing them this season. Those who dared to wear them get bonus points for the risk-taking. Though something tells us they'll be as obsolete as a Nintendo 64 come next season.

This chapter looks at AI-based marketing and advertising. The example of MSCHF is just a case-in-point of how brands, the Internet, and AI are becoming intertwined. The company's bizarre-looking boots, which look like something out of a video game, became a fad first among celebrities, who were seen wearing them conspicuously in public, thus making them highly popular among the general public subsequently. This event showcased, in microcosm, the synergy that has always existed between advertising, marketing, celebrity culture, and pop culture more generally, but in a new strange way. Ads and commercials have always been devised in the past as era-specific texts, reflecting the social codes of the different eras in which they were ensconced. For example, Coca-Cola's "Universal Brotherhood" campaign in the early 1970s, directly incorporated the images, rhetoric, and symbolism of the counterculture movement into its representational style and content. With the emblematic jingle "I'd like to teach the world to sing in perfect harmony," the campaign showed the extent to which cultural movements, advertising, and marketing had become interrelated.

It is difficult to see how and where the MSHC boots campaign fits into this synergistic system (Gao and Zhao 2021), other than being just a useless fad, much like the pet rock fad of the 1970s, worth revisiting here retrospectively. The pet rock was introduced in 1975 by advertising entrepreneur Gary Dahl in Los Gatos, California. Dahl claimed that he created the rock to cater to the many people he met who complained about their animal pets, which had to be bathed, groomed, walked, brought to the vet, and so on. A pet rock, on the other hand, was easy to keep as a companion. The rock came with a 32-page "training manual" which included instructions on how to properly care for it. Commands such as "sit," "stand," and the like were accompanied by pictures of how to nudge the rock appropriately. Incredibly, the pet rock caught on as a huge fad. Dahl appeared on the *Tonight Show* and by Christmas of 1975, two and a half tons of the rocks were sold, with newspapers, radio programs, and other television shows featuring segments on the fad. Copycat rocks flooded the market shortly after. However, by 1976, the pet rock had become "last year's fad."

Various theories have been put forward to account for the popularity of the rock. Whatever the reason, the episode of the pet rock is a source of insight into the MSCHF boots fad, which is strikingly similar. The main difference is that the boots were designed by an AI, and advertised online, indicating that the world of advertising and marketing had migrated to the world of hyperreal marketing. Brands, the Internet, and AI are becoming intertwined indistinguishably in the promotion and representation of brand products.

Advertising and Pop Culture

Significantly, to enhance the desirability of its products MSCHF presented itself as offering unique viral marketing opportunities for celebrities, even creating shoes designed for various musicians to help boost their album streams. The MSCHF example, however, represents an old pattern, with a key update—the synergy between advertising and pop culture has been projected onto the world of AI and the Internet (as mentioned above). MSCHF also tapped into another traditional strategy of marketing—creating controversial products that would spark public debate, thus garnering a lot of opportunistic media attention. This kind of stratagem is spreading across the brand world and its many online promotion schemes. Another kind of strategy involves retrieving the techniques of the pop art movement. Some products themselves are designed as pop art objects, including the MSCHF red boots, which could easily have been devised by a Warhol or a Rauschenberg. As Kordic (2015) aptly puts it: "In an age where digital technology completely took over our lives, Pop Art comes in as a given, with characteristics that match our current environment perfectly."

To reiterate here (Chapter 6), a common technique for the pop artists was depicting brand-name commercial products, with their sculptures of hamburgers and other fast-food items. They also appropriated the techniques of commercial (assembly-line) mass production, abandoning individual, titled paintings in favor of large series of works, all depicting the same objects. A classic example is Andy Warhol's series of silkscreens of the Campbell's soup can in 1962. At first, Warhol's cans were exact replicas of the actual cans, but as he continued with the series, the images became more and more surrealistic. The soup company was unimpressed and even upset at this portrayal of their iconic product. Eventually, it realized that the Warhol silkscreens had engrained the can into American folklore. In 2004, the company released a limited-edition series of the cans that were inspired by Warhol's paintings. Pop art emerged, in essence, to redefine the roles and techniques of art in a consumerist economy. The paradoxical relation between pop art and consumerism constitutes a core feature of the whole pop culture paradigm, given that, as Fatima Sumair (2022) observes, "Appropriating images from mainstream culture and putting them on canvas tends to change our relationship with the items as viewing something as 'art'," which then urges us "to become more present with the image" and thus "to think more critically about it."

The partnership between advertising and pop culture goes back to, at least, the latter years of the nineteenth century, when advertising started mirroring the new cultural trends and fads in the creation of the ads themselves. The British Pears Soap company was among the first to understand the power of advertising to induce a favorite image of its products, embedding messages into its ads that connected the product to some social trend. In one of its famous campaigns, it showed a young child and a puppy happily together, thus painting a positive portrait of childhood and domestic animals, both of which required soap (literally and metaphorically) to maintain not only cleanliness but also a sense of ideal living. The first of its ground-breaking marketing campaigns used the exact same image in a well-known statue by Italian artist Giovanni Focardi, titled *You Dirty Boy!*, exhibited at the Exposition Universelle de Paris in 1878, showing a mother scrubbing her young son with soap. The statue was so popular that Pears purchased the rights to use it as part of its marketing campaigns. Pears even made celebrities of some of the personages it employed for its ads. For example, British actress Lillie Langtry became the first woman to endorse a commercial product—Pears soap.

It was obvious from the outset that the objective was to influence people to perceive objects of consumption as necessary accouterments of life, leading to a widespread need for new things, which Roland Barthes (1967) famously called *neomania*, which he actually claimed had originated in the Renaissance, but became embedded as a mass psychological syndrome in consumerist culture. As the cultural styles changed so too did advertising accordingly, the two reflecting and influencing each other, imprinting into groupthink the perception that consumption and culture were intertwined in everyday life. The Frankfurt School philosopher, Herbert Marcuse (1964: 123), criticized this phenomenon cynically as follows: "If mass communications blend together harmoniously, and often unnoticeably, art, politics, religion, and philosophy with commercials, they bring these realms of culture to their common denominator—the commodity form. The music of the soul is also the music of salesmanship. Exchange value, not truth value, counts."

It was radio that further embedded the culture-advertising partnership into social groupthink by introducing the so-called commercial—a mini-narrative or musical jingle revolving around a product or service and its uses. The commercial became itself a form of entertainment, and since it could reach masses of potential customers it raised advertising to a *sui generis* subgenre of pop culture itself. Early commercials, such as *Folger* coffee's pseudoscientific sales pitches, *Mum Deodorant's* satires of spy movies, and *Pepsodent* toothpaste's snappy jingles, became so familiar across the nation that perception of the products became inextricably intertwined with the styles and contents of the commercials created to promote them. In other words, pop culture, advertising, the marketplace, and technology formed a mutually-beneficial alliance in the 1920s, remaining an enduring one to this day. As Castells (2000) has cogently suggested, this alliance has been both instrumental and adaptive, shifting its *modus operandi* to the digital universe and, now, to the AI world.

With AI-generated brands such MSCHF, the synergy between advertising and pop culture has further obliterated the distinction between the two. The ability of AI to segment customers, message them through social media platforms at optimal times, and to personalize ad messages according to statistically-assessed tastes (Kaput 2022), makes it a "master advertiser." Recall the deepfake video produced by the Spanish fashion house Balenciaga, which reconstructs the cast of the Harry Potter saga as a satirical commercial for the brand (Chapter 1). In the comments found on social media sites with regard to the video, it would appear that the two narratives—the original and the commercialized one—were perceived as indistinguishable. As Zeeshan Ahmad (2023) asks, given this audience reaction: "What is the difference [between] Potter and Balenciaga?" The answer is provided sardonically by the video itself at the end as follows: "There is no good and evil. There is only Balenciaga and those too weak to seek it."

Clearly, in the AI Age, advertising has adapted, but still maintains its intrinsic interrelationship with pop culture trends. An example was the AI-powered ad campaign to promote the highly successful 2023 *Barbie* movie. A notable element of the campaign was the "Barbiefication" selfie generator, which allowed users to upload a photo of themselves that would be transformed into a Barbie-like image. The Barbie brand itself used AI to create personalized content for fans, such as AI-powered quizzes that allowed users to learn more about their own "Barbie tastes," and even conceiving some of the quiz content for older audiences, so as to evoke a sense of nostalgia for the doll, given that it has been a part of many people's childhoods, even though the movie was a parody of the original cultural meanings of the doll as representing the female childhood experience.

AI Ads

Advertising is the psychological fuel that drives consumerism. Print ads and audio-visual commercials gain appeal, momentarily, because they are designed as artistic, creative texts, tapping into existing social codes. Often, they are so effective that they are adopted and incorporated into various domains of pop culture. In the 1950s, for example, the Silly Putty, Slinkies, and Hula Hoops toy products became so popular that they were the inspiration for songs. Silly Putty was introduced in 1949 by advertising marketer Peter C. L. Hodgson, who discovered a substance developed by General Electric researchers looking for a viable synthetic rubber. The silicone substance could be molded like soft clay, stretched like taffy, and bounced like a rubber ball. Slinky was a coil toy that could be made to "walk" down a staircase by itself by placing it on a higher step in a specific way. The Hula Hoop was a light plastic hoop that could be whirled around the body for play or exercise by the movement of the hips. These brands became integrated with pop culture, remaining so to this day, because they were promoted effectively through creative socially-coded advertising. The subtext in the early ad texts was that of fun for the whole family, focusing on the view of childhood as a phase involving play with toys—tapping into a widespread social code related to the rearing of children. The cultural salience of the Hula Hoop was even acknowledged by Banksy with a 2020 street art painting in Nottingham, England, which shows a young girl using a bicycle tire as a Hula Hoop.

The strategy of coopting contemporary social codes to create ad texts is now even more evident in the training of AI on large databases that connect cultural themes with consumerist trends (David 2023). It is thus little wonder to find that marketers have adopted generative AI *en masse*, using it to help them write effective and trendy ads that tap into the dominant concerns, trends, and even crazes of everyday life. The key advantage of AI-generated ads is, in fact, that they can be built around themes that are considered important by society and can be changed adaptively, via ongoing algorithmic assessments, as these themes change (Dean 2022; Rogoff 2023). If, for instance, there is a broad satirical mood in online pop culture trends, then the AI picks up on it and produces ads that correlate with the mood, including the manufacturing of "satirical products," such as MSCHF's red boots. Needless to say, satire is not a new technique in the advertising world. But in the world of algorithms, the type of satire that is coopted by the deep learning systems is a "present" one—in effect, the satire used in previous ads, if continued in the present, would likely have little or no effect; the satire that is evident in the online world, on the other hand, can be easily harnessed by the algorithm to create ad texts that reflect its spirit. MSCHF understood, above all else, that we have been conditioned by the pop art movement to conceive of advertising as art, adopting of marketing strategies and ads that satirize our consumeristic lifestyles, at the same time that they induce them—reflecting the age-old paradox of advertising. MSCHF understood that poking fun at everything is now a common mood, as can be seen in everyday banal social media communications.

Many brands have integrated AI to tap into social moods in the same way as MSCHF. Coca-Cola even launched a competition-oriented campaign called "Create Real Magic," where they invited consumers to combine ChatGPT, DALL-E, and historical Coca-Cola ad texts to create new ones that would be shared on the company's website—reflecting the use of digital tools for self-styled creativity. Nike combined AR (augmented reality) with AI, allowing users to scan their own feet with an app that measures exact shoe sizes, which can then be passed on to the company, thus personalizing a product via technology. AR brings elements of the virtual world into the real world, thus enhancing sensory perception, from hearing to seeing. Nike used AI to create an ad campaign called "Never done evolving," based on an cloned imagery that pits a younger version on Serena Williams with a more contemporary one, thus tapping into the deepfake trend that garners a lot of attention and "Likes," as discussed several times in this book. In another AI-based campaign, Nutella enlisted AI to generate seven million unique labels for jars of the chocolate. The jars sold extremely well in the marketplace, perhaps because consumers may have felt that they were purchasing a one-of-a-kind product.

Interestingly, a trend has emerged that seems to make fun of advertising itself, using AI tools, mirroring the so-called subvertisements of the culture jamming movement, which were spoofs or parodies of actual ads and ad campaigns (Lasn 2000). A case-in-point was a 2023 video ad about a made-up franchise called Pepperoni Hug Spot (Fig. 7.1).

The tagline was generated by ChatGPT, the images by MidJourney, the video by Runway Gen2, the voiceover by Eleven Labs, and the music by Soundraw AI. The video ad was put together by an anonymous person on Reddit. The fact that it was about a fake pizza eatery did not seem to rattle anyone on social media; rather, the video went viral for a time, with

Fig. 7.1 AI-Created ad for a fictional pizza (from Twitter, @Pizza_Later)

comments on social media sites indicating that people saw it as funny, since it featured quasi-photorealistic people eating the pizza in hilarious ways, slurping it down as if it were a liquid. Another video that similarly went viral in the same time frame featured a deepfake version of actor Will Smith devouring loads of spaghetti over and over (Fig. 7.2).

What was the aim of the video? Was it meant to be satirically funny or just weird? The creator used Stable Diffusion to make it. The fact that it too, like the pizza video, went viral suggests that there is an online trend that has taken its cue from the culture jamming movement, as mentioned, which emerged in the early 2000s. Through their website, culture jammers provide satirical critiques of advertising, with clever satirical ads. The idea is that once viewers take a critical look at the mimicked ad campaign, they are forced out of their emotional comfort zone, impelling them to think about the implications presented by the jammer. Is this the aim of those who make the satirical AI videos? It is difficult to actually make a connection between such ads and the culture jamming movement, given that satire and parody seem to constitute a kind of discourse in online representations and communications. If nothing else, it is clear that the AI tools that are now at everyone's disposal have allowed for a do-it-yourself system of ad textuality that attracts many at first and quickly subsides into the ether of cyberspace.

Fig. 7.2 Deepfake Will Smith video (Screenshot from YouTube)

To use a notion put forth by Castells (1996), AI-generated culture, from ads to movies, is an "ever-present" one, in which, as Benjamin (1927: 23) cynically observed in the first decades of the twentieth century, "history breaks down in images not into stories."

Overall, AI is now an intrinsic tool in ad creation (and its parodies), requiring less and less human involvement. As Rogoff (2023) warns: "This would give the advertising industry a new generation of automated systems with novel capabilities and untold impacts, and deepen the dangers that digital platforms pose to our society." The fact that deepfake and authentic advertising evoke the same pattern of receptivity by the denizens of the online world is indirect evidence of the veracity of this statement. Indeed, as Rogoff (2023) goes on to assert, the business world itself welcomes the deepfake and personalized ad spoof trend because it further embeds advertising as an unconscious *langue* that plays on the power of images, like the MSCHF marketing stratagems conspicuously demonstrate:

> These risks go beyond ads themselves: The business model encourages the design of platforms where inflammatory or controversial user-generated content (not just ads) spreads faster, because this content attracts eyeballs, which can then be redirected to ads.

Hyperreal Marketing

The ways in which MSCHF marketed its products, via AI advertising, is a perfect example of "hyperreal marketing," defined as marketing using AI-based synthetic media and showcasing products on the computer screen or via immersive Virtual Reality-Augmented Reality systems. As the MSCHF collective realized, the world of the Matrix and life in hyperreality has congealed broadly, including in consumerist culture. In hyperreality, consumers can interact with the brand and with celebrity users. They can even design their own versions of a brand. Hyperreal marketing has in effect tapped into the ever-expanding desire to eliminate all traditional boundaries, including those that have kept consumers and brand designers and makers apart as separate entities. As Neo put it in *The Matrix* (1999), the hyperreal world is attractive because it is a "world without rules or controls, borders or boundaries. A world where anything is possible. Where we go from there is a choice I leave to you" (wikiquote.org/wiki/The_Matrix_(film)). It is little wonder that hyperreal marketing is expanding, given our propensity to assume that we are in control of everything, by simply clicking a mouse, touching a screen, or voicing a command to an AI personal assistant. Like MSCHF, Roblox is a company that has made hyperreal marketing its core strategy. It is an online game platform and game creation system, which allows users to design their own games and to play games created by other users. Roblox had over 200 million users in 2021.

Other companies have followed the MSCHF and Roblox examples. In 2022, Coca-Cola released a beverage called Sugar Byte, with ads that were targeted at gamers, who are a core demographic in the hyperreal universe. The company promoted the drink with the appropriate slogan, the “taste of pixels.” Of course, the drink tasted like any other Coca-Cola product, but the metaphor tapped into the hyperreal mindset of consumerist culture. The company even created an entire (hyperreal) island called “Pixel Point in Fortnite,” where players could interact with the virtual version of Sugar Byte as well as play different sensory-based games. Even clothiers are providing hyperreal experiences, which unfold via apps. In 2019, the iconic traditional clothier, Gucci, started selling shoes by giving buyers a chance to “try on” those very shoes via an app. The potential customer points a smartphone downward to the feet and then chooses a shoe style icon, from which the app generates a digital image of the shoes overlaid on the customer’s feet. The app can track foot movements, and so customers can look at the virtual shoes in real time from different angles. The idea is to provide a sense that customers are making the decision to buy the shoes on their own, not through any persuasive interaction with a human seller or as a consequence of some persuasive advertising campaign.

Hyperreal marketing is a microcosmic reflection of what a world increasingly run by AI implies, where the human and the machine are no longer distinguishable as agents of creativity. Some implications were already enunciated by Baudrillard in 1981, who foresaw a world where simulation and dissimulation were no longer seen as falsehoods, but as (desirable) options to truth and reality He put it prophetically as follows (Baudrillard 1981: 3):

> To dissimulate is to pretend not to have what one has. To simulate is to feign to have what one doesn’t have. One implies a presence, the other an absence. But it is more complicated than that because simulating is

> not pretending: Whoever fakes an illness can simply stay in bed and make everyone believe he is ill. Whoever simulates an illness produces in himself some of the symptoms. Therefore, pretending, or dissimulating, leaves the principle of reality intact: the difference is always clear, it is simply masked, whereas simulation threatens the difference between the "true" and the "false," the "real" and the "imaginary."

As Baudrillard (1996: 23) subsequently put it, focusing more pertinently on advertising: "We will never know if an advertisement or opinion poll has had a real influence on individual or collective wills, but we will never know either what would have happened if there had been no opinion poll or advertisement." As the MSCHF case brought out conspicuously, hyperreality is, as Baudrillard anticipated, an apt framework for explaining the psychology of current consumerism, which emphasizes a simulation of individualism. The core premise in hyperreal marketing is, in fact, that satisfaction or happiness is achieved by individuals who believe that they are in control of their lives and even other people's lives, a mindset achieved through simulation. But this is an AI-generated mythology. As Nick Land (2022) has observed, reality engenders superstitions, while hyperreality engenders "hyperstitions," which are superstitions induced via simulations that bring about their own perceived reality.

The Frankfurt School Redux

The Frankfurt School, as discussed previously (Chapter 4), caustically attacked what they saw as the commodification of culture, whereby cultural products (musical, artistic, literary, etc.) are treated as manufactured products, produced massively, and quickly discarded shortly thereafter. While their critiques have been challenged on various counts (as discussed previously), which ignore the fact that pop culture has produced truly remarkable works of art, one wonders what the School would say about AI-generated ads and hyperreal marketing campaigns, given that they saw cultural production as being controlled by advertising and commercial imperatives, thus serving to create subservience to the system of consumerism itself. As HorkheimerAdorno (1944: 23) put it in the *Dialectic of Enlightenment*:

> Advertising and the culture industry merge technically as well as economically. In both cases the same thing can be seen in innumerable places, and the mechanical repetition of the same culture product has come to be the same as that of the propaganda slogan. In both cases the insistent demand for effectiveness makes technology into psycho-technology, into a procedure for manipulating men. In both cases the standards are the striking yet familiar, the easy yet catchy, the skillful yet simple; the object is to overpower the customer, who is conceived as absent-minded or resistant.

Adorno (1970) maintained that the commodification of culture was controlled, in fact, by those who held financial power. Jürgen Habermas (1987) similarly claimed, decades later, that mass cultures have inculcated the view that the pursuit of pleasure through consumption was life's overarching goal. As AI-based marketing continues to spread and evolve, perhaps the Frankfurt School might need to be revisited for insights, as Matthew Handelman (2022) has written: "Revisiting the Frankfurt School in the age of algorithmic reason shows that, contrary to views foundational to computing, a neural-network chatbot does not sidestep meaning but rather carries and alters it, with unforeseen social and political consequences." In contrast to mass produced culture, including deepfake and other forms of AI-generated culture, true art cannot be reproduced so easily and is not subject to algorithmic combinatorics. It is also not subject to marketplace economics, nor to satirical appropriations of advertising, as is the case with a company such as MSCHF. Despite the many counter-arguments to Adorno's views, they seem to have a special resonance today, strangely and ironically.

When the Internet came into wide use, it was heralded as bringing about a liberation from conformity and a channel for expressing one's opinions freely. But this view has proven to be specious. In the same way that the Frankfurt School scholars critiqued the Fordist world of mass production culture in the pre-Internet era, it can be said that simulated AI-generated culture is built on the same promise of the attainment of individual agency through artificial means. Living in a social media universe, we may indeed feel that it is the only option available to us. The triumph of the Internet, especially in the AI Age, lies in its implicit promise to allow human needs to be expressed individualistically, yet connect them communally—hence the paradox.

Epilogue

AI is an ever-expanding and powerful "culture-producing machine," generating texts of all kinds, regardless of their artistic or aesthetic value. Included in the textual mix are AI-generated ads that are increasingly catapulting consumers into the world of hyperreality, where they can gain a sense of self-styled control over the products themselves, much like the kind of control that gamers believe they have over gaming conditions and outcomes. In an intriguing article, John Herrman (2023) asks the type of question that pertains to AI in advertising: What happens when ads generate themselves? Superficially one might simply answer that the time has arrived for hyperreal marketing in the marketplace, which is just a new form of marketing. But this indifference evades the larger question that such marketing implies for the future of a society that has evolved according to the artistic canons of pop culture, and which has been in a partnership with the marketplace since at least the 1920s. Are we witnessing the end of this partnership or, on the contrary, are we beholding its solidification in hyperreality? As Herrman (2023) remarks, one thing is clear: AI ads and marketing strategies are effective because they blend in with the cultural ecology in which we live:

> These AI ad-generation products are less visible to regular users than, for example, an Instagram bot pretending to be Tom Brady or a pop-up asking if you'd like help responding to an email. But they're also clearer in purpose and tell us a lot about what sorts of problems these companies think generative AI can solve — for them, anyway. Take Amazon, which recently started using AI to summarize reviews and to let sellers automatically generate most of a listing—"compelling product titles, bullet points, and descriptions"—from a few words.

With AI-generated hyperreal locales, such as the ones created by MSCHF, Roboblox, Coca-Cola, and others, we may actually be experiencing a new kind of consumerist consciousness shaped by simulation—mirroring the broader simulacrum state of mind, as Baudrillard (1983) warned. Such companies have in effect tapped into the simulacrum, further entrenching it, showing concretely how the world is morphing via AI technologies. As mentioned, MSCHF creates controversial products and projects that generate a lot of attention, blending elements of art, commerce, and technology, that impel us to, paradoxically, see their products as *au courant* at the same time that they are critiques (subvertisements) of the products themselves. MSCHF even created a single-edition sculpture titled "The Persistence of Chaos," which was a laptop infected with six of the most infamous computer viruses in history. It sold at auction for over a million dollars. Was it a symbolic statement about the world in which we live, where dependency on computers is a form of Orwellian Big Brotherhood? Was it a way to comment on the absurdity of consumerism itself, in true Andy Warhol style?

Perhaps a general answer to such questions is that MSCHF and other hyperreal brands have understood today's mood and even ethos, shaped by social media and other virtual worlds, such as a video games. Like many people today, MSCHF resolutely refuses any kind of identifying consumerist label, with the employees even hesitant to even call themselves a "company." MSCHF thus puts forth a new version of authenticity, perhaps the only kind that can be achieved by some form of indirect consensus in the world of the Matrix—a world that is based on simulation, rather than actual reality. As Luca Silipo (2019) has aptly observed, we are becoming entangled more and more in the solipsism of Selfhood, entrenched by the worlds of the Internet and social media, which have destroyed traditional views of authenticity:

> Modern society appears more and more entrenched in an ideal of 'one'. The ethics of authenticity is trivialised into a narcissistic and solipsist self-referentiality…There are 'ones' everywhere. But 'ones' is simply not what makes our society sustainable…Behind their façade of online sociality, Internet, social media, and user-generated content platforms have been misused as beauty contests where the exaltation of the selves compete with each other by the most audacious selfie, the most disrespectful Insta-challenge, or the loudest influencer.

References

Adorno, Theodor (1970). *Aesthetic Theory*. London: Bloomsbury.

Ahmad, Zeeshan (2023). Harry Potter by Balenciaga: Is This the AI-Generated Future of Entertainment? *T-Magazine*, https://www.tribune.com.pk/story/2409540/harry-potter-by-balenciaga-is-this-the-ai-generated-future-of-entertainment.

Allaire, Christian (2023). Why Is Everyone Wearing These Cartoonish Red Boots? *Vogue*, https://www.vogue.com/article/everyone-is-wearing-mschf-big-red-boots.

Barthes, Roland (1967). *The Fashion System*. Berkeley: University of California Press.

Baudrillard, Jean (1981). *Simulacra and Simulations*. New York: Semiotext(e).

Baudrillard, Jean (1983). *Simulations*. New York: Semiotexte.

Baudrillard, Jean (1996). *The Perfect Crime*. New York: Verso Books.

Benjamin, Walter (1927). *Arcades Project*. Cambridge: Harvard University Press.

Castells, Manuel (1996). *The Rise of the Network Society: The Information Age: Economy, Society, and Culture*. New York: Wiley-Blackwell.

Castells, Manuel (2000). *The Rise of the Network Society*. Oxford: Blackwell.

David, Emile (2023). Big Companies Use AI-Generated Ads Because They're Cheap. *The Verge*, https://www.theverge.com/2023/8/18/23837273/generative-ai-advertising-oreos-cadbury-watermarking.

Dean, Ian (2022). This AI-Generated Advertising Is Creeping Me Out. *Creative Bloq*, https://www.creativebloq.com/news/advertising-ai.

Gao, Yin and Zhao, Lin (2021). Synergistic Effect Between Online Broadcast Media and Interactive Media on Purchase Intention. *Frontiers in Psychology* 12, https://doi.org/10.3389/fpsyg.2021.781743.

Habermas, Jürgen (1987). *The Philosophical Discourses of Modernity*. Cambridge: MIT Press.

Handelman, Matthew (2022). Artificial Antisemitism: Critical Theory in the Age of Datafication. *Critical Inquiry* 48, https://doi.org/10.1086/717306.

Herrman, John (2023). What Happens When Ads Generate Themselves? *Intelligencer*, https://nymag.com/intelligencer/2023/10/what-happens-when-amazon-and-meta-ads-generate-themselves.html.

Horkheimer, Max and Adorno, Theodor (1944). *Dialectic of Enlightenment: Philosophical Fragments*. Stanford: Stanford University Press.

Kaput, Mike (2022). AI in Advertising. *Marketing Artificial Intelligence Institute*, https://www.marketingaiinstitute.com/blog/ai-in-advertising.

Kordic, Angie (2015). What Is Pop Art in the 21st Century? *Widewalls*, https://www.widewalls.ch/magazine/what-is-pop-art.

Land, Nick (2022). *The Dark Enlightenment*. Imperium Press.

Lasn, Kalle (2000). *Culture Jam: The Uncooling of America*. New York: Morrow.

Marcuse, Herbert (1964). *One-Dimensional Man*. London: Routledge.

Rogoff, Zak (2023). Generative AI's Overlooked Risk: Ads. *Democracy: A Journal of Ideas*, https://democracyjournal.org/arguments/generative-ais-overlooked-risk-ads/.

Silipo, Luca (2019). "The Ethics of Authenticity" by Charles Taylor Is an Important Book. *Medium*, https://luca-silipo.medium.com/the-ethics-of-authenticity-by-charles-taylor-is-an-important-book-5be32b973302.

Sumair, Fatima (2022). Pop Art and Technology. *Medium*, https://medium.com/%40fatimasumair/pop-art-and-technology-af99be7d5381.

CHAPTER 8

AI and Gaming Culture

There is a definite argument to be made that videogames are becoming an art form put together by artists of different types.
—John Romero

Prologue

A sci-fi-based video game called *No Man's Sky* was launched in 2016 as a virtually endless game, using a technique called "procedurally generated content" to create simulated fauna, flora, animals, topography, and atmospheric conditions for its facsimile planets, of which there were 18 quintillion unique variations. The game was created with algorithms that generated its format, layout, avatars, and content based on data inputted by the game developer. It showed, for arguably the first time in gaming culture, that AI had the potential to assign real creative control to players, given that it also provided players with the ability to deviate in randomized ways from the game's initially-programmed narrative structure. In effect, the game was its own world, into which players entered as its denizens with the ability and freedom to control elements and events in that world, with their own ingenuity and resourcefulness.

No Man's Sky was an AI-generated simulated world. Its narrative was about survival and exploration of unknown universes, where players can fly spaceships, search for new physical and chemical elements, and then

M. Danesi, *AI-Generated Popular Culture*,
https://doi.org/10.1007/978-3-031-54752-2_8

use them to build structures and survive in new non-Terra environments (Santos et al. 2023). The game offered, in fact, an abundant amount of creative freedom to players to plan and act as they charted and explored a fictional world. In its own microcosmic way, the game provided a concrete framework for discussing not only the general implications it harbored of incorporating AI into any domain of pop culture, but also a concrete format for considering the so-called *simulation hypothesis* (Bostrom 2003), which claims that we are all living in an extremely powerful computer program, in which we are ourselves constructs. This chapter will look at AI-generated video gaming, in fact, as another domain for assessing how pop culture itself may be evolving or vanishing, as we have known it for over a century. Video games are authorless video texts that put narrativity in the hands of the players, literally and metaphorically, with avatar characters that act and behave via control mechanisms that affect how the plot unfolds. AI has even made "non-player characters" (NPCs)—simulated characters in a game not controlled by a human player of the game—a common feature of video gaming, further blurring the perceptual boundaries between simulation and reality (Virk 2019).

AI has actually been an integral part of video games since their inception in the 1950s, but the AI technology at the time was a rudimentary one, when compared to current technologies. Game playing was, as a matter of fact, a key area of investigation within AI itself from the start, as discussed previously, with games such as chess and checkers constituting field laboratories for research into AI systems. Role-playing games that featured a single player mode with avatar enemies started appearing in the 1970s; and by the 1990s, AI made immersion and interactivity routine. With the advent of generative AI, the complexity and variety of the narrative textuality of video games are truly unbelievable, as are the realistic and natural voices and gestures of the NPCs that match the tone and mood of their speech. By using data about players gathered from previous gameplay, generative AI can now create unique storylines and tailor-made quests.

Video Games and Popular Culture

Games played with electronic devices on a screen go back to the early era of pop culture, reflecting an inherent use of technology for play and diversion. The roots of the modern technology-based entertainment industry were arguably implanted in the 1880s, when coin-operated game contraptions started appearing in train depots, hotel lobbies, and restaurants, eventually developing into pinball machines in the early penny arcades, whereby players would score points by directing the path of a metal ball on an artificial playfield in a glass-covered case. Pinball gaming expanded considerably after World War II with the addition of several new mechanical devices within the game. In 1948, the first completely electronic pinball machine, called the "Cathode Ray Tube Amusement Device," was introduced into the arcades, becoming the first veritable video game. As Jeremy Saucier (2016) has noted, one cannot underestimate the role that the early pinball machines played in both mirroring and shaping "broader changes in social and cultural values, art, popular culture, game design, technology, and politics."

With computer technology expanding in the 1970s, the first video game adapted for television, called Odyssey, became a household item virtually across America. In the same era, a computer engineer named Nolan Bushnell founded a video game development company, called Atari, which produced *Pong*, a simple tennis-style game that became widely popular on the first personal computers in the late 1970s, an era when video gaming became a huge business, remaining so to this day, especially after the arrival of *Pac-Man* in 1980, where simulated characters exhibited different personality traits, adding psychological interest to gaming. In the millennium, video game consoles such as Xbox and PlayStation became major sources of entertainment, emblemizing how technology has guided pop culture trends, moods, and even art since at least the 1920s (Muriel and Crawford 2018).

One of the most compelling features of even the earliest role-playing video games (RPGs) is that they provide a format for imaginary adventure which is narrativized by the players themselves "on the go," so to speak, not by some pre-established narrator or text, although the RPG's narrative theme may come from some preexisting storyline or mythology. RPGs became entrenched into the pop culture landscape with *Dungeons & Dragons*, released in 1974. It was the first modern RPG, establishing many of the conventions that have come to characterize the genre, simulating the fantasy worlds of the traditional adventure story, but leaving the actual plot and outcome up to the players, who imagine themselves to be knights, wizards, or some other mythological avatar possessing great physical and moral strength. The spreading popularity of the RPG was highlighted shortly thereafter by the 1982 made-for-television movie *Mazes and Monsters*, which was about a group of college students and their obsessive interest in RPGs.

As Gary Fine (1983) noted at the start of the RPG craze, the central attractive feature of the games was in fact that they allowed players to assume and enact attributes that they did not necessarily possess in reality, such as physical strength, social poise, good looks, and other attractive qualities. In RPGs, the players are the scriptwriters, actors, and directors at once. Whatever their effects on the mind, which have been studied abundantly since they became popular (Johnson 2005; Doan and Strickland 2012; Kriss 2020), the relevant point here is that RPGs have been an intrinsic part of the overall engaging quality of pop culture itself since they emerged as pinball machines (Hawreliak 2018). With AI-enhanced games such as *No Man's Sky*, they can thus be used in theory as gauges of how pop culture might itself be evolving (or mutating) with the advent of generative AI, including the increased sense of empowerment and personalization of textuality that AI permits. As Hogeweg (2021) comments: "It's very easy for a *No Man's Sky* player to choose a corner of the universe and make it entirely their own, diligently building up whatever they want for their own enjoyment without worrying about coming across other players."

A plausible psycho-evolutionary reason for why such games have intrinsic appeal might be the presence of a ludic instinct in the human species, evidenced by the fact that competitive games actually go back to prehistory (Olivastro 1993: 5–11). Dutch historian Johan Huizinga (1938) even characterized the human species as descending from *homo ludens*, to emphasize the role of play in the constitution of early cultures. A key inference of Huizinga's theory is that the structure of human thought might be fundamentally ludic, constituting a creative force in the formation of human understanding and reflecting a need for order within what would otherwise be perceived as randomness. As Friedrich Hayek (1988: 194) has put it:

> The practices that led to the formation of the spontaneous order have much in common with rules observed in playing a game. To attempt to trace the origin of competition in play would lead us too far astray, but we can learn much from the masterly and revealing analysis of the role of play in the evolution of culture.

Indirect support for Huizinga's theory comes from the fact that there has never been a period of time, nor has there ever been a culture, without some kind of ludic tradition. Interestingly, in RPG culture, one of Huizinga's key concepts, which he called the "magic circle," has been incorporated into the general structure of the game's *situs*. This is conceived as the space in which the normal rules of reality are suspended and replaced by the artificial rules of the game world. As Castronova (2005: 147–148) has noted, this space "can be considered a shield of sorts, protecting the fantasy world from the outside world," but which will also exhibit characteristics of the outside world, because it "cannot be sealed completely [since] people are crossing it all the time in both directions, carrying their behavioral assumptions and attitudes with them." The magic circle of an RPG is, essentially, where the narrative crystallizes (Salen and Zimmerman 2003). Huizinga (1938: 10) described the magic circle as follows:

> All play moves and has its being within a play-ground marked off beforehand either materially or ideally, deliberately or as a matter of course. Just as there is no formal difference between play and ritual, so the 'consecrated spot' cannot be formally distinguished from the play-ground. The arena, the card-table, the magic circle, the temple, the stage, the screen, the tennis court, the court of justice, etc., are all in form and function playgrounds, i.e., forbidden spots, isolated, hedged round, hallowed, within which special rules obtain. All are temporary worlds within the ordinary world, dedicated to the performance of an act apart.

The distinction between a character or action that occurs in the game and a person or action outside of the game is becoming increasingly inconsequential to players, as the RPG technology is becoming increasingly a generator of the simulacrum effect. But this does not implode reality, as Baudrillard (1981) believed, but rather provides a version of it that assigns control to players. Actions taken within the games succeed or fail according to a formal system of rules and guidelines, which each player utilizes. Actually, as Kathleen McClancy (2018) notes, RPGs constitute an ideal medium for assessing the validity of simulacrum theory itself, given that the "ubiquity of visual simulations, their self-referentiality, and their inescapability seem to be proving Baudrillard's warnings about the descent of postmodernity into the 'desert of the real'." However, as she goes on to note, "the loose narratives of the games create a pervasive atmosphere of distrust of technology that leads players to see through both the embrace of simulacra and the control of rule-systems" (McClancy 2018)—an observation that can be applied to the paradox of pop culture in general as both a culture providing entertainment and the source of deep reflections on the nature and state of the modern world.

AI Video Games

One of the first AI-based RPGs was *Assassin's Creed*, first released in 2007, an open-world RPG that revolves around a fictional millennia-old struggle between the Order of Assassins, who physically fight for peace and freedom, and the Knights Templar, who desire peace through control. The game featured historical characters intertwined with real-world events and figures. The players start playing the game as "Assassin Initiates." In subsequent editions, with improved AI, new stories and time epochs are introduced. The editions follow Desmond Miles, a descendant of several historical Assassins who uses a machine called the Animus to retrieve the past in the present.

The concept of the magic circle is illustrated perfectly by the "Atlas" in *No Man's Sky*, where it is depicted as a floating diamond-shaped structure where a computer system stores all possible data, reminiscent of the Wizard of Oz narrative. Once inside the Atlas, the player is confronted by a giant, shape-shifting orb, which promises true understanding of reality. The path for reaching the Atlas, however, presents the player with challenges and insights at the same time, much like the Wizard of Oz storyline; the difference is that guidance along the path is provided by an alien monolith, recalling the monolith symbol in *2001: A Space Odyssey* (1968), which represented the epic transitions in human evolution from ape-like beings to civilized people and then space travelers. The *No Man's Sky* monolith symbolizes a new path for evolution, one in which humans become integrated with machines. However, there is another subtext in the game that stands out, which can be seen in the part of the game when, after accepting the monolith's guidance, the player comes to an awareness that humans may be nothing more than lines of code, created by the computer in the Atlas.

The game thus provides a simulated immersive environment for grasping the simulation hypothesis (discussed below) concretely, through the game form itself—a kind of virtual "hands-on" process of understanding. A furtherance of this mindset can be seen in subsequent

AI-generated RPGs, which have come to use AI to endow individual players to assume the power of wizards (Stokel-Walker 2023). Semiotically, such games have replaced real-world referents by their simulated representations, which become psychologically real through the game. The notion that applies specifically in this case is that of iconicity, introduced by Charles Peirce (1931–1958) as a semiotic process of transferring by resemblance perceivable properties of the world into sign forms, as for instance, constructing words that carry forth within them the imitation (simulations) of the sounds perceived in their referents (*splash*, *bang*, *crunch*, etc.). Simulative iconicity is intrinsic to video games, making them essentially huge iconicity machines.

However, since the iconicity occurs in hyperreality, itself a simulation, players are projected into a simulation of a simulation, where realism and fantasy mesh completely, as the referents of the iconic process exist only in the game space, where users create cities, landscapes, people, adventures, and even ideas, which, as Myk Eff (2022) has perceptively remarked, become "the space in which the game takes place, and the rules and conventions of the game dictate the behavior of the players within that space." Perhaps in no other medium, can people engage with iconically-created sign structures, which are hard to leave semiotically behind when re-entering the real world (Le et al. 2023). In fact, as Courtney Borrett (2022) has aptly observed, AI video gaming has induced many to envision the game world as one in which a "strange cultural zeitgeist has resulted in the emergence of parasocial relationships." Significantly, Xiang (2023) describes a research project at Google and Stanford that used ChatGPT to generate human-like characters who live and interact in a simulated world called Smallville. The characters sleep, make breakfast, interact with each other, and engage in behavior that mimics human behavior as closely as possible. The same AI is being used to create more interesting NPCs in video games, making the games even more compelling to players.

Artificial intelligence has been making significant strides in recent years, with applications ranging from healthcare to transportation. One area where AI has been making a particularly strong impact is, as emphasized here, in the world of video games. Game developers are increasingly using AI to create more realistic and immersive experiences for players, as Xiang has also noted. However, as AI-enhanced video games become more sophisticated, psychological and ethical questions inevitably arise concerning such aspects as the following: (a) How does the illusory representation of reality that the video game creates affect perceptual and cognitive processes in players? (b) How can game developers ensure that the boundary between reality and hyperreality is clearly delineated in terms of the game itself?, One can never ignore the potential consequences of virtual experiences on players. As AI algorithms become more advanced and more capable of generating increasingly realistic environments and characters, the simulacrum effect may become so powerful that "players may struggle to distinguish them from real people, leading to emotional attachments and potential psychological consequences," as Marcin Frąckiewicz (2023) observes. If AI-generated video games become a diffuse form of entertainment for a larger and larger portion of the population, this could have enormous implications for pop culture: Will all art forms, such as music and film, be transferred to the video game world? Will avatars become the new performers in this world? It is no coincidental happenstance event that AI is now being used to create music for video games, as well as art backgrounds.

The Simulation Hypothesis

The key "discovery" that players make in *No Man's Sky*, as mentioned, is that humans might be nothing more than computer code, and thus may actually be living in a world constructed by a computer system. This encapsulates, in essence, the so-called simulation hypothesis (SH), which was introduced as such, not by a scientific or philosophical treatise, but first by a movie, *The Matrix* (1999), and then a spate of subsequent movies dealing with the same hypothesis, such as *eXistenZ* (1999), which is about a video game designer, named Allegra Geller, who has created a game called "eXistenZ", a game in which players have "bioports" attached to their backs through a hole drilled into the spine, and a pod device that hooks directly into that hole, allowing the game system to access the user's memories, desires, wants, and fears. But the devices are not entirely synthetic; they are strange living, breathing things. After an attempt on her life by a crazed fan of the game, Allegra goes on the run with Ted, a young businessman. In an attempt to save her game, and thus "existence," Allegra implants the pod that carries a copy of eXistenZ into Ted's body.

The simulation hypothesis implies that everything we see around us, our universe, the galaxies, planets, and life forms, is a methodically programmed computer simulation. The physical laws governing reality are algorithms and the feelings and experiences we have are generated by computational processes. The key point here is that the SH has gained traction, not because it was researched and debated mainly in academic and scientific circles, but because of pop culture—that is, via films and video games. The following lines recited by Morpheus in *The Matrix* encapsulate the hypothesis (https://www.imdb.com/title/tt0133093/quotes/):

> —What is real? How do you define 'real'? If you're talking about what you can feel, what you can smell, what you can taste and see, then 'real' is simply electrical signals interpreted by your brain.
>
> —You have to let it all go, Neo. Fear, doubt, and disbelief. Free your mind.
>
> —Have you ever had a dream, Neo, that you were so sure was real? What if you were unable to wake from that dream? How would you know the difference between the dream world and the real world?

The 2022 Netflix series, *1899*, was another filmic treatment on the SH. The narrative centers on the character Maura Franklin, a neurologist who finds herself aboard a ship carrying migrants hoping to start a new life in America. Mysterious events occur throughout the story, which are finally explained in its last few episodes, revealing that the ship is a simulation that has been manipulating the passengers' reality. Everything from their memories to their physical appearance is a construct of the simulation, which actually consists of an infinite number of layers of simulations running inside other simulations, also ad infinitum.

It was philosopher Nick Bostrom (2003) who formulated the SH in academic terms. Since antiquity, philosophers have wondered if there are levels of reality. A classic example is Plato's Allegory of the Cave in his *Republic* (375 BCE). Plato imagined several prisoners trapped in a cave underground. They are chained against one wall, while on another they see shadows of objects that are projected from the outside world by a large torch held behind their heads. The prisoners only know the reality of the cave, and so mistake the shadows they see on the wall as part of reality itself. Eventually, one of the prisoners escapes and sees the real objects outside, realizing the nature of the deception. In *The Matrix*, this is framed as "red-pilling," introduced in a scene where Morpheus offers Neo (then a real-world hacker named Thomas Anderson) a choice between a blue or red pill. The former would allow Anderson to return to his normal life, unaware of the Matrix, whereas the red pill would lead him to enlightenment—awareness of the fact that reality is a computer simulation.

Bostrom posits the existence for a technologically advanced civilization that possesses immense computing power but needs only a fraction of that power to simulate realities and construct beings in them. Several interesting studies came forth after Bostrom's proposal to provide an investigational basis to the SH. One study (Campbell et al. 2017: 78) tested the SH "based on the assumption that the system performing the simulation is finite (i.e., has limited resources)." To achieve the computational level, such a system would, as in a video game, have to be able to "render content (reality) only at the moment that information becomes available for observation by a player and not at the moment of detection by a machine (that would be part of the simulation and whose detection would also be part of the internal computation performed by the Virtual Reality server before rendering content to the player)" (Campbell et al. 2017: 78). Outside of a video game there is no evidence that this can be achieved.

The movie *Ready Player One* (2018), set in 2045, imagines a world in which much of humanity has access to a system that allows people to escape the real world, an obvious analogy to Plato's prisoner who was able to escape the cave. Interestingly, the simulated world to which people go is indistinguishable from the real world they leave behind. Perhaps this is an allusion to the SH itself as unverifiable; moreover, the whole movie has the look and feel of an RPG, further emphasizing the fact that pop culture, rather than academia, is where the SH is concretely being debated. Part hyperstition, part imaginary computer science, part excitement over the potential of AI, the SH would likely not have sparked the modern imagination without *The Matrix* and *No Man's Sky*. It is an idea that, by its very existence, has reified itself.

AI-Generated Narrative Content

RPGs based on AI systems provide opportunities for the creation and development of narrativity within the game format. As Myk Eff (2023) has emphasized, narrativity is at the core of the game system, but in AI-generated game content, it is hardly a linear storyline that the game allows, with the usual arc of a beginning, a middle, and an end; rather, the narrative is in the hands (literally) of the players, making it aleatory, albeit within a conceptual frame defined by the genre of the game and the algorithms involved in creating it. Such open narrativity is what allows players to engage interactively and non-linearly with the game world, with its characters, topographical features, rules, and even moral laws.

RPGs fall into two categories, historically, which can be called "closed" and "open" games, to adopt a distinction introduced by Umberto Eco (1989) with respect to traditional narrative texts. Before the advent of sophisticated AI in game design, RPGs were designed to permit a relatively small set of varying actions and movements. They were "closed games," called progression games, since the sequence of events is constrained by the structure of the game. The classic example is the original *Super Mario Bros* game released in 1985 by Nintendo. Players controlled the characters of Mario, his brother Luigi, and other family members in a multiplayer mode. The main plot involved getting the brothers to traverse the so-called Mushroom Kingdom to rescue the Princess Peach from an evil character. The journey was made possible by side-scrolling mechanisms and the avoidance of obstacles and enemies with the aid of power-ups. The game made possible very few variations to this plot structure. On the other hand, a game such as *No Man's Sky* is an "open game," because the plot and outcome are not clearcut, but open to various alternative storylines that AI-generated content makes possible, including different quests, characters, dialogues, and other elements that are tailored to a player's unique experience. Open games are self-organizing structures, and as such constitute ludic models of what has come to be called "emergentism" in science and philosophy (Soler-Adillon 2019). This is the theory that a complex entity has properties that its parts do not have on their own, emerging via their interaction in a wider whole. For this reason, they are also called emergent games.

Whether a game is a closed or open one, the narrative that it enfolds retains elements of traditional narrative themes, such as those identified by Vladimir Propp (1928), who argued that there are a small number of such themes, which he called *narremes*, that undergird every story, from fairytales to novels, such as *good*, *evil*, *journey*, *combat*, etc. While a game such as *No Man's Sky* may seem to have no semblance of a traditional story, it still has narremic structure, with the themes of a journey (into the unknown), of combat with alien enemies, and so on. It is this unconscious narremic system in open RPGs that produces the sense of being "lost" in the fictional worlds they provide to players, made ever more realistic with cutscenes, or the non-interactive sequences that interrupt the game-play, so that "conversations" between characters can occur. According to Eric Hayot (2021: 180), RPGs actually indicate how narrative has evolved from antiquity to today, as people become more enmeshed personally in the story because of technology, while in the past it was only through paraphrase and explanation that a story could be personalized.

As one last comment on RPGs, it is relevant to mention the so-called Proteus Effect, since it brings out the cognitive and emotional power that RPGs have on some players. This is the phenomenon that emerges when the behavior of individual players is influenced by the characteristics of their avatar. The roles that players adopt, and the plot, narrative, and virtual environments in which they immerse themselves, provide glances into life experiences different than their own (Yee and Bailenson 2007). Having interactions with other characters in a game while playing as avatars can grant players the opportunity to assume the identity of someone else and act differently than how they normally would. However, this is not unlike the imitation of fictional characters, such as comic book superheroes, that children tend to do, becoming, temporarily, like the heroes they admire.

Epilogue

The following citation by Stone (2019) can be used to summarize, synoptically, the overall theme of this chapter:

> Games now allow players to take control of the narrative, by giving them choices with meaningful consequences that affect the outcome of the story. Storytelling has matured, going from meaningless damsel in distress cases that serve to motivate the player, to being at the forefront of the game with real messages and important stories to tell.

As in other domains of pop culture, generative AI has clearly entered the world of video games changing it in significant ways (Marr 2023). It is noteworthy that video games have been used as models for discussing both the simulation hypothesis and even the theory of emergentism, as discussed above. The ways in which RPGs are suggestive of philosophical themes is actually a new element in their history, given that they started out as mere entertainment devices; so too is the fact that they can become so powerful psychologically that players may not want to escape from them. The question of the psychological effects of RPGs on its users has become ever more relevant. As Gary Fine (1983: 4) observed already in the 1980s, for many players, the games might constitute the main reality, because they may be "willing to bracket their natural selves and enact a fantasy self."

Leaving aside the debates on such effects, video gaming can be seen to have a broad appeal because it provides a means of creating imaginary worlds autonomously, unlike the fictional worlds created by reading a novel, watching a movie, and so on. The RPGs take the make-believe narrative element away from a single author and put it directly into the hands of the player. As a consequence, the RPG world has become a virtual *sui generis* pop culture, with its own stories, music, celebrities (avatars), and so on. As Taylor (2006) has argued, the games provide an imaginary locus in which players can live out the fantasy as adventurers, heroes, and explorers, enticing them to create their own form of story-book culture, apart from real-world culture (Adams and Smith 2008).

The advent of AI-created game culture dovetailed with the movie *The Matrix*, which showed what life in a simulated virtual world entailed psychologically. It is therefore not surprising to find that video games based on the *Matrix* set of movies have become highly popular, projecting players into the world of the Matrix both literally (the game itself) and metaphorically (the mind-world that the movies portray). The overall model of life as a video game actually imbues all the movies with a philosophical element. As Neo puts it in *The Matrix Resurrections* (2021), encapsulating the overall thematic subtext of the movies: "They took my life and turned it into a video game." In the film, Neo is again trapped in the simulated world which he had tried so hard to leave in the original movie. But this time, he is a successful game designer himself who has created a highly successful and famous trilogy of video games with the same plot, characters, and stylistic elements as the previous Matrix films. The original *Matrix* portrayed the virtual world as a hyperrealistic space that can be hacked to paradoxically expose its unreality, from which Neo wanted to escape. But in *The Matrix Resurrections* the inhabitants of that space seem not to care that they are immersed in it. Maybe this is a sign of how the world is morphing. As Robertson (2021) remarks in his review of the film: "people understand they're being fooled and just don't care. They like the comfort of fondly remembering old stories at the expense of anything new or real. They experience things the way someone has told them to experience them, even when there's nothing actually there." Perhaps, the powerful attraction to video games is that they allow for complete immersion into the world of simulation, where people can gain some control over things, at least in an artificial way. As James Burns (2022) has aptly observed:

> Games are the ultimate Matrix. They are artificial worlds that we willingly enter through the proxy of a controller. They become as good as real thanks to our deliberate suspension of disbelief.

References

Adams, Tyrone L. and Smith, Stephen A. (eds.) (2008). *Electronic Tribes: The Virtual Worlds of Geeks, Gamers, Shamans, and Scammers*. Austin: University of Texas Press.

Baudrillard, Jean (1981). *Simulacra and Simulations*. New York: Semiotext(e).

Borrett, Courtney (2022). Pushing the Boundaries of AI and Psychology in Darkwebstreamer. *Game Developer*, https://gamedeveloper.com/design/pushing-the-boundaries-of-human-psychology-procedural-generation-and-artificial-intelligence-in-darkwebstreamer.

Bostrom, Nick (2003). Are You Living in a Computer Simulation? *Philosophical Quarterly* 53: 243–255.

Burns, James (2022). It's Time for a New Matrix Video Game. *Medium*, https://medium.com/super-jump/its-time-for-a-new-matrix-video-game-574752a234a3.

Campbell, Tom, Owhadi, Houman, Sauvageau, Joe, and Watkinson, David (2017). On Testing the Simulation Theory. *International Journal of Quantum Foundations* 3: 78–99.

Castronova, Edward (2005). *Synthetic Worlds: The Business and Culture of Online Games*. Chicago: University of Chicago Press.

Doan, Andrew P. and Strickland, Brooke (2012). *Hooked on Games: The Lure and Cost of Video Game and Internet Addiction*. FEP International.

Eco, Umberto (1989). *The Open Work*. Cambridge, MA: Harvard University Press.

Eff, Myk (2022). The Magic Circle, https://soundand.design/the-magic-circle-febdba265c22.

Eff, Myk (2023). Narrative Elements, https://soundand.design/narrative-elements-e91d2fab863e.

Fine, Gary Alan (1983). *Shared Fantasy: Role-Playing Games as Social Worlds*. Chicago: University of Chicago Press.

Frąckiewicz, Marcin (2023). Ethical Implications of AI-Enhanced Video Games: Reality, Representation, and Responsibility. *TS2 Space*, https://ts2.space/en/ethical-implications-of-ai-enhanced-video-games-reality-representation-and-responsibility.

Hawreliak, Jason (2018). *Multimodal Semiotics and Rhetoric in Videogames*. New York: Routledge.

Hayek, Friedrich (1988). *The Fatal Conceit*. Chicago: University of Chicago Press.

Hayot, Eric (2021). Video Games & the Novel. *Daedalus* 150: 178–187.

Hogeweg, Stan (2021). No Man's Sky's Total Number of Planets explained. *Gamerant*, https://gamerant.com/no-mans-sky-how-many-planets-explained/.

Huizinga, Johan (1938). *Homo Ludens: A Study of the Play-Element in Human Culture*. New York: Beacon Press.

Johnson, Steven (2005). *Everything Bad Is Good for You: How Today's Popular Culture Is Actually Making Us Smarter*. New York: Riverside.

Kriss, Alexander (2020). *The Gaming Mind: A New Psychology of Videogames and the Power of Play*. The Experiment.

Le, Kim, Ortiz de Gortari, Angelica, Callan, Annabel, Poynton, Daragh, Vecchio, Daniela, and Chen, Wai (2023). Game Transfer Phenomena in a Clinical Case with Psychosis and Gaming Disorder. *Psychiatry International* 4: 286–296.

Marr, Bernard (2023). Generative AI Is Breathing New Life Into Classic Computer Games. *Forbes*, https://forbes.com/sites/bernardmarr/2023/09/22/generative-ai-is-breathing-new-life-into-classic-computer-games/.

Muriel, Daniel and Crawford, Garry (2018). *Video Games as Culture*. London: Routledge.

McClancy, Kathleen (2018). The Wasteland of the Real: Nostalgia and Simulacra in Fallout. *Game Studies* 18, https://gamestudies.org/1802/articles/mcclancy.

Olivastro, Dominic (1993). *Ancient Puzzles: Classic Brainteasers and Other Timeless Mathematical Games of the Last 10 Centuries*. New York: Bantam.

Peirce, Charles S. (1931–1958) *Collected Papers of Charles Sanders Peirce*, Vols. 1–8, C. Hartshorne and P. Weiss (eds.). Cambridge, MA: Harvard University Press.

Plato (375 BCE). *The Republic*, Internet Archives, https://archive.org/details/a604578400platuoft/.

Propp, Vladimir J. (1928). *Morphology of the Folktale*. Austin: University of Texas Press.

Robertson, Adi (2021). The Matrix Resurrections Has an Incredibly Smart Take on Showing Video Games in Movies. *The Verge*, https://www.theverge.com/2021/12/23/22850518/matrix-resurrections-video-game-movie-simulation-reality.

Salen, Katie and Zimmerman, Eric (2003). *Rules of Play: Game Design Fundamentals*. Cambridge: MIT Press.

Santos, Diogo, Zagalo, Nelson, and Morais, Carla (2023). Players Perception of the Chemistry in the Video Game No Man's Sky. *Simulation & Gaming* 54, https://journals.sagepub.com/doi/10.1177/10468781231169301.

Saucier, Jeremy (2016). Pinball Matters. *Museum of Play*, https://www.museumofplay.org/blog/pinball-matters.

Soler-Adillon, Joan (2019). The Open, the Closed and the Emergent: Theorizing Emergence for Videogame Studies. *Game Studies* 19, https://gamestudies.org/1902/articles/soleradillon.

Stokel-Walker, Chris (2023). AI Generates Video Game Levels and Characters from Text Prompts, *New Scientist*, https://newscientist.com/article/2390019-ai-generates-video-game-levels-and-characters-from-text-prompts/.

Stone, Chris (2019). The Evolution of Video Games as a Storytelling Medium, and the Role of Narrative in Modern Games. *Game Developer*, https://www.gamedeveloper.com/design/the-evolution-of-video-games-as-a-storytelling-medium-and-the-role-of-narrative-in-modern-games.

Taylor, T. L. (2006). *Play Between Worlds: Exploring Online Culture*. Cambridge: MIT Press.

Virk, Rizwan (2019). *The Simulation Hypothesis*. Bayview Books.

Xiang, Chloe (2023). Google Tells AI Agents to Behave Like 'Believable Humans' to Create 'Artificial Society'. *Vice*, https://vice.com/en/article/z3mvj3/google-tells-ai-agents-to-behave-like-believable-humans-to-create-artificial-society.

Yee, Nick and Bailenson, Jeremy (2007). The Proteus Effect: The Effect of Transformed Self-Representation on Behavior. *Human Communication Research* 33: 271–290.

CHAPTER 9

Simulacrum Culture

Our brain simulates reality. So, our everyday experiences are a form of dreaming, which is to say, they are mental models, simulations, not the things they appear to be.
—Stephen LaBerge

Prologue

In 2016, a Japanese YouTuber named Kizuna attracted millions of views and followers. But the "personage" was not a human being, but rather an avatar generated with computer graphics. Thousands of virtual YouTubers have since populated social network platforms, garnering huge audiences, with viewers interacting spontaneously with them as if they were real human beings. Along with increasingly sophisticated technologies, such as Virtual Reality (VR), the Kizuna phenomenon can be seen to constitute a concrete sign of the ever-broadening spread of a "simulacrum culture," whereby engagements with virtual characters, settings, and so on are no longer seen as oddities, but as alternatives to human-based interactions in the real world. The seamless blending of AI and VR technologies is not only revolutionizing the ways we do important things in society, such as train pilots, doctors, make effective diagnoses of diseases, model urban planning more efficiently, and so on, but also how we perceive and consume entertainment. Baudrillard's (1981) apprehension

M. Danesi, *AI-Generated Popular Culture*,
https://doi.org/10.1007/978-3-031-54752-2_9

that a simulacrum culture would eventually come about to the detriment of humanity has occurred, perhaps not as detrimentally as Baudrillard had assumed, but rather as a new form of pop culture, which (not coincidentally) had been predicted by the sci-fi literature (Chapter 1).

Hollywood started producing movies with early VR-based computer graphic technology already in the 1970s, with films such as *Star Wars*, *Terminator*, and *Jurassic Park*. Video games also started adopting these technologies at around the same time frame. Since then, VR has blended with AI to increasingly provide entertainment alongside human-only spectacles, or in coordination with them, as in VR-based projections of external audiences into a live concert. This has rather significant implications for the future of pop culture itself, which may even be replaced by a simulacrum culture, in which textual representations and avatars will tend to be accepted in themselves as legitimate elements in entertainment and even artistic engagement, as discussed in the previous chapter with regards to video games. If that occurs extensively, then "the map" (the representation) would have become indistinguishable from "the territory" (what it stands for), to adopt a famous phrase by the Polish-American scholar Alfred Korzybski (1933: 58)—a phenomenon whereby the simulation is meaningful in itself. Alluding indirectly to Korzybski's notion, Baudrillard (1981: 1) made the following apt observation:

> Today abstraction is no longer that of the map, the double, the mirror, or the concept. Simulation is no longer that of a territory, a referential being or substance. It is the generation by models of a real without origin or reality: A hyperreal. The territory no longer precedes the map, nor does it survive it. It is nevertheless the map that precedes the territory—precession of simulacra—that engenders the territory.

Baudrillard (1981) also made a key distinction between *simulacra* and *simulations*—the former are copies of things that either have no original referentiality, or whose meaning has dissipated through constant reproduction, whereas the latter are straightforward imitations of real-world things. The two are often intertwined. VR itself is both a "simulacrum machine," immersing us into a world that does not actually exist, but it is at the same time a simulated world that is designed to resemble the real world. After a while of being immersed in a VR world, the difference between the two is lost, as the experience tends to become a true simulacrum.

This chapter will deal the notion of simulacrum culture, given that more and more pop culture texts and spectacles are actually simulacra. It will also look at the issue of deepfakes of artists and celebrities, which are no longer seen as illegal or even dishonorable, but as legitimate simulated forms in the pop culture world. Finally, it will discuss the Metaverse, which is the hyperreal space that links VR with AI technology, and which has become a new locus for fiction, recalling, strangely, the origin of fiction in the medieval era as a literary art.

Simulacrum Theory

The term *simulacrum* derives from Latin, meaning "likeness" or "similarity." It appeared for the first time (as far as can be told) in the Latin poem *De Rerum Natura* by Lucretius (c. 99 BCE). It was used in the nineteenth century by painters to describe drawings that were accurate copies of other paintings, rather than emulations of them. Baudrillard (1981) recycled the term, as a representation that is a copy of something made up, and thus non-existent in reality. According to Baudrillard, there are four levels of simulative representation, with simulacra consisting the fourth level.

1. *Reflection of reality*. This is a representation that is designed to refer some aspect of reality. A common example is a portrait painting that attempts to reproduce the anatomy and facial structure of the actual subject of the portrait, as in early Renaissance painting.
2. *Perversion of reality*. The term perversion, as can be gleaned from Baudrillard's description, indicates an alteration of meaning, which could also involve a distortion. For instance, a Surrealist painting is a representation that hints at reality through a juxtaposition of images in reality as they might be imagined to occur in a dream state.
3. *Pretense of reality*. This type of representation pretends (aims) to refer to something that actually does not exist in the real world. Mythic images, such as the unicorn, can be seen to occur at this level.
4. *The simulacrum*. This is a representation that is a copy of something that does not exist, but which comes into existence via the representation. It has been extended to include the creation of mind worlds, as in video games, which are simulations, but eventually become more real than real in the imagination. Literary critic Fredric

> Jameson (2007) famously offered photorealism as an example of a simulacrum, in which a painting is created by copying a photograph that is itself a copy of the real thing.

Aware of the relevance of Baudrillard's notion to the ever-broadening world of AI-generated culture, *The Matrix* (1999) was the first film to make specific allusions to Baudrillard's notion of the simulacrum. At the start of the movie, Neo is seen holding a copy of Baudrillard's 1981 book, *Simulacra and Simulations*, which he hollows out so as to create a secret hiding place for cash and computer files. Neo's hollowed copy shows the chapter "On Nihilism" in the middle, not at the end of the book, where it is in Baudrillard's actual book, perhaps intimating that we are now right into the middle of the simulacrum. In another episode, Morpheus refers to the real world outside the Matrix as the "desert of the real," a specific reference to Baudrillard's work—a phrase he utters after Neo wakes up from his computer-generated Virtual Reality, experiencing the "Real" as a desolate yet spectacular territory. In fact, the simulacrum world of the Matrix gains its power because of its ability to envelop people from the real world into its simulations of reality, as in a video game. The simulacrum thus surpasses the experience of the real world, becoming hyperreal, which experienced as more-real-than-real. Like the main protagonist, Neo, people now tend to experience reality "on" and "through" the computer screen, and consciousness is largely shaped by the simulacrum that the screen generates. Only in such a mind-world could a virtual YouTuber such as Kizona attract such large audiences willing to interact with it.

It is interesting to note that Gilles Deleuze (1968) saw simulacra instead as important social constructs because, in his view, they challenged "privileged positions" of authorship and are thus transgressive of rigid cultural paradigms. Umberto Eco (1990: 23), on the other hand, saw simulacra as leading to the substitution of what is real with imitations in the collective imagination: "we not only enjoy a perfect imitation, we also enjoy the conviction that imitation has reached its apex and afterwards reality will always be inferior to it." In a simulacrum environment, not only is cultural-artistic meaning (as we have experienced it traditionally) fundamentally challenged, but so is our everyday perception of what is original or authentic and what is just an imitation. As Baudrillard warned, it is this indifference that will eventually project humans completely into hyperreality. As Nosta (2023) affirms in agreement with Baudrillard's ominous prevision: "In this age of AI-enabled hyperreality, it's imperative that we stay connected to the underlying truths of our world, ensuring that the map doesn't replace the territory."

As Dave Hallmon (2023) has perceptively commented, Baudrillard's theory provides "a profound metaphor for our current cultural shift toward desiring the hyper-reality provided by Generative AI and social media...As we become increasingly immersed in these simulations, it becomes more and more difficult to discern what is real and what is artificial." This constitutes what can be called a "hyperreal culture syndrome" that manifests itself in everything from the popularity of a Kizuna YouTuber to the abandonment of literature as a mirror of the world or of painting as a unique human enterprise, as discussed in previous chapters. From 2016 to 2021, Kizuna was the most subscribed influencer on YouTube. Because Kizuna primarily communicated in Japanese, fans even contributed translations for her videos, and there were fan clubs built around these translators, until 2020 when official subtitles began to appear. The avatar became a veritable celebrity, performing at virtual concerts, appearing on television shows, and starring in television commercials. This could only have happened, as Baudrillard maintained, in a world of hyperreal culture, even before the advent of generative AI. The dangers of hyperreality are not in seeking meaning in simulations, but, as Baudrillard (1988: 64) maintained, but in seeking it constantly in them, whereby "we are gorged with meaning and it is killing us."

Virtual Reality

The 2018 film *Ready Player One* revolves around Oasis (Ontologically Anthropocentric Sensory Immersive Simulation), a Virtual Reality simulation game, which allows people to escape the real world. Control of the game goes to the person who can track down an Easter egg hidden inside the game, which is a massively multiplayer online simulation game. The film's subtext seems to be that the world is moving more and more into hyperreality, as simulacrum culture becomes ever more a fact of everyday life via the possibilities that generative AI makes possible. The choice of the acronym OASIS is a suggestive one, since the word "oasis" connotes a peaceful area or period in the midst of a difficult, troubled, or hectic situation. Could this mean that video games are an oasis, a symbol of utopia and fulfillment? On the other hand, could the term be a subtle allusion to Baudrillard's "desert of the real," since an oasis is a spot in the desert where water is found? A key aspect of OASIS is that the larger it becomes, the more it encourages users to spend all their time in it. Is this a foreshadowing of the spread of simulacrum culture in lieu of any other form of culture?

The technology that makes the OASIS compelling is a merger of AI and VR. The latter term, which refers to an environment designed to render the feeling of being immersed in a real world, is, as mentioned above, a simulacrum of reality. The origin of the term is traced to 1938 when French playwright, Antonin Artaud called the verisimilitude to real life of characters and objects in the theater as engendering in the minds of audiences a "*réalité virtuelle*." The English term was introduced in a 1982 science fiction novel titled, *The Judas Mandala*, by Damien Broderick. Its technical designation is due to Jaron Lanier, who designed the first VR hardware in the same era of the 1980s (see Lanier 2017). In 1960, before Lanier's work, cinematographer Morton Heilig designed a machine, called "Sensorama," that was designed to stimulate a movie audience's senses artificially. It consisted of a chair that could tilt, allowing the viewer to stare at a wide-angle screen, which showed three-dimensional films accompanied by sound and odor emitters. Heilig was unable to get funding to develop his machine, but it nonetheless prompted research into subsequent VR technology. In 1961, the Philco Corporation developed the "Headsight"—a video screen put on a head mount, linked to cameras, which produced a viewing angle when users turned their heads. In 1965, Ivan Sutherland created a graphics accelerator for his "Ultimate Display" device, connected to a computer, which displayed graphics and allowed head movements to change the view. In 1989, the sensory glove was invented to allow users to experience real hand movements in virtual spaces.

VR technology is powerful psychologically in producing a simulacrum effect because it can transmit the sights, sounds, and sensations of simulated reality to the user who may experience them as actually real. VR also has the ability to record and send the speech and movements of the participants to the simulation program, which further enhances the simulacrum effect, projecting the human subject into a simulated representational space where the user is interacting with the representation. To see in the virtual world, the user wears a head-mounted display (HMD), which contains a position tracker to monitor the location of the user's head and the direction in which the user is looking. Using this information, a computer recalculates images of the virtual world to match the direction in which the user is looking and displays these images on the HMD. Users hear sounds through earphones in the HMD. Studies have consistently shown that, after a while, users tend to believe that they are in a real environment, or that the distinction between the real and the simulation is irrelevant, as the user becomes sensorially involved increasingly in the VR space (Lavoie et al. 2021).

VR technologies have been changing not only the way we interact in video games, but also the way we experience movies and other popular media. Nintendo's Wii is an example of how VR devices can easily become a part of everyday life. Wii makes it possible for individuals to play golf or tennis, or to engage in boxing, in the simulacrum environment it generates. VR systems are clearly on the cusp of changing people's perceptions of what representations are and what role they play in everyday experiences. They can also be used for beneficial purposes, not just entertainment. A research project by Stanford University's Virtual Human Interaction Lab found that if someone is immersed in a VR world and observes a tree being cut down, feeling the vibration and hearing the chainsaw and tree crashing to the ground, that person is more likely to conserve paper subsequently (Gorlick 2011). It is the ability of VR to project the viewer imaginatively into an environment, such as the one just described, which induces the person to experience the danger as if it were real.

Another relevant project (Ahn et al. 2016) has suggested that immersive digital environments can stimulate a positive sense of connection with Nature, raising awareness that could potentially make humans more sensitive to the dangers of climate change. Specifically, the project aimed to determine if VR would get users to empathize with the imaginable experiences of animals in Nature. In one experiment, the researchers used VR to get subjects to "enter the body" of a cow, crawling on hands and knees and stooping in a bovine way to eat and to drink water, just before they were prodded into a vehicle and transported to a slaughterhouse. Another experiment used VR to simulate coral dying in an acidifying ocean, with participants feeling the vibrations and hearing the cracking sounds of the coral's branches breaking off. The researchers showed the same two scenarios on video to a control group. They found that the VR participants empathized with the creatures in scenarios more than the control group. Beyond its potential uses as an instrument for changing people's minds about the dangers we are facing today, as the above research implied, the relevant point to be emphasized here is that VR has entered the pop culture domain in several concrete ways. Live concerts are now filmed and streamed in VR, allowing fans of a musician or band to virtually put themselves in the middle of the crowd at a concert. With various cameras set up to capture 360-degree images from a variety of angles, including on the stage, the concerts can reach a wider audience in a more reality-simulating medium than on a standard television screen.

Perhaps the most salient use of VR technology is, as mentioned several times, in gaming culture. A VR game allows players full immersion into the simulated world of the game. For example, the so-called Oculus Rift, considered to be the first consumer-useable VR headset for video games, released in 2016, was incorporated into video game systems broadly. The device allows for 360-degree views and realistic sounds, creating a sense of physical presence within the virtual world. In no other simulated environment does the simulacrum effect become so engaging as in such games. Pop culture and simulacrum culture may soon become synonyms, as mentioned. Of course, pop culture has always projected people into the fictional worlds of superheroes, adventures, idyllic romances, and the like. But audiences largely observed these worlds in a detached way. With VR, the detachment is ceding to immersion.

Deepfake Culture

One of the most famous of all paintings is Leonardo da Vinci's *Mona Lisa* (*La Gioconda* in Italian). Below is the renowned image (Fig. 9.1):

However, it is not a screenshot of da Vinci's masterpiece; it is a deepfake generated by AI. This is not an isolated or exceptional event. Deepfakes of artworks, celebrities, performers, and the like have become intrinsic elements of a broadening simulacrum culture (Lanz 2023). Recall the fact that fans of the real Drake and The Weeknd performing the AI-generated song "Heart On My Sleeve" (Chapter 1) actually liked the deepfake voices, constituting just one type of concrete evidence that the simulacrum effect is now an unconscious mindset.

The term "deepfake" emerged in 2017 when a Reddit moderator created a subreddit called "deepfakes," posting videos that used face-swapping technology to insert the likeness of celebrities into erotic videos. Deepfakes have become worrisome on many fronts, since they can be used illegally to deceive people and thus manipulate them. Deepfake technology, as the "Heart On My Sleeve" example conspicuously brought out, has evolved to be increasingly convincing. Consider so-called Unreal

Fig. 9.1 Deepfake *Mona Lisa* (Midjourney, Wikimedia Commons) (Info: A rendition of the Mona Lisa by Midjourney AI. Created with the prompt "Oil painting of the Mona Lisa by Leonardo DaVinci": https://en.wikipedia.org/wiki/File:Mona_Lisa_by_Midjourney_AI.jpg: (Attribution: "*This file is in the public domain because it is the work of a computer algorithm or artificial intelligence and does not contain sufficient human authorship to support a copyright claim.*")

Keanu Reeves, a popular TikTok personage, who is a deepfake recreation of the actor, which received millions of views. The fake Keanu Reeves dresses in everything from sweat pants to expensive-looking suits, and typically talks about his movie roles, and often discusses etiquette in the current age. The account provides a disclaimer—"Parody, Life and Eternal Youth." But this did not seem to matter to most of the followers, a symptom of the spread of simulacrum culture. As Schwartz (2023) found in surveying comments on Reddit, there were those who truly thought it was Reeves ("Why wouldn't this be him?" "It looks like him to me"), those who thought it was a lookalike ("I don't understand how you're not Keanu. Separated at birth? Lol. It's just too uncanny"), and those who detected the fakery ("Obviously fake, maybe Unreal Engine?"). Despite the fact that the account name made it saliently clear that the video was about an "unreal" Keanu Reeves, it appears that it does not matter to most of the followers.

Examples such as the Unreal Keanu Reeves have spread throughout social media, including Deep Voodoo's deepfake video for Kendrick Lamar's "The Heart Part 5" (Deep Voodoo is an AI entertainment company), in which the rapper's face gradually metamorphizes into the face of Kanye West one moment and the fact of Will Smith the next. Deep Voodoo has created many celebrity deepfakes, including those of the actors of the 1977 film, *Star Wars*. As Baudrillard (1981: 97) wrote, at some point the engagement with hyperreality leads to the condition whereby "People lose the ability to distinguish between reality and fantasy. They also begin to engage with the fantasy without realizing what it really is."

The advent and growing entrenchment of a deepfake culture, a spin-off of simulacrum culture, leads to a revisitation and updating of Walter Benjamin's epochal 1934 study, "The Work of Art in the Age of Mechanical Reproduction" (mentioned previously), in which he maintained that "even the most perfect reproduction of a work of art is lacking in one element: Its presence in time and space, its unique existence at the place where it happens to be." So, a painting captured as a photograph or a digital image may be able to reproduce the look of the painting down to the fine details, but the reproduction will lack the unique feeling that the real painting evokes by viewing it directly. The aura, as discussed, is thus lessened in the reproduction of an artwork. The concept is somewhat equivalent to Karl Bühler's (1934) notion of *energeia* in communication, which he envisioned as a kind of "spiritual (non-physical)" sense we get from semiotic artifacts such as words and paintings—constituting examples of *ergon*, a conceptualization that goes back to Aristotle (385 BCE). A deepfake may, however, possess a kind of fake aura, which seems to fit in with simulacrum culture producing its own kind of synthetic aesthetics. It might mean, on the other hand, that the notion of aura belongs to a different era, and may now have suffered the same "death" as that of the author. Already in 1964, Paul Valéry made the following relevant statement, which rings more true today than it did then (Valéry 1964: 225):

> Our fine arts were developed, their types and uses were established, in times very different from the present, by men whose power of action upon things was insignificant in comparison with ours. But the amazing growth of our techniques, the adaptability and precision they have attained, the ideas and habits they are creating, make it a certainty that profound changes are impending in the ancient craft of the Beautiful. In all the arts there is a physical component which can no longer be considered or treated as it used to be, which cannot remain unaffected by our modern knowledge and power. For the last twenty years neither matter nor space nor time has been what it was from time immemorial. We must expect great innovations to transform the entire technique of the arts, thereby affecting artistic invention itself and perhaps even bringing about an amazing change in our very notion of art.

As Cover (2022: 609) has insightfully remarked, deepfake culture is not an aberration; it is the outcome of the evolution of pop culture itself as based on fiction and on "widespread shifts in the cultural sense of texts, representation, play, co-creativity and pastiche."

The Metaverse and the Narraverse

Simulacrum culture has found an online locus—the Metaverse, where users can interact with a computer-generated environment and the avatars of other users. The term was first used by Neal Stephenson in his 1992 novel *Snow Crash*. As a VR environment, the Metaverse is accessed with the same kind of technology discussed above for video games.

The Metaverse is an adjunct to both simulacrum and traditional pop culture. An example is the Motown label and its artists who now exhibit themselves on STYNGR, which has millions of tracks (Brandle 2023). STYNGR is an online platform offering an in-game music experience with popular tracks for the gamers, who can explore the label's music and catalogue, as well as attaching music snippets to their avatars. In April of 2022, the Metaverse's first fashion week debuted, featuring fashion enthusiasts, designers, and clothing brands in a VR space where activities mirrored real-life fashion events , including catwalks, musical afterparties, talkbacks, and the like. Two years earlier, in April of 2020, the video game Fortnite featured a virtual event starring an avatar of rapper Travis Scott, which attracted nearly 30 million players. Shortly after, Roblox featured Lil Nas X in a digital performance attracting over 33 million.

In the Metaverse, the simulacrum becomes an extension of everyday life, not a faux replacement or alternative to it. In some ways, the Metaverse parallels the advent of fiction in the medieval era, as an imaginary medium to escape reality—a reality shaped by the bubonic plague at the time. Literary critics, in fact, peg Giovanni Boccaccio's *Decameron* (1348) as the moment in which prose fiction as a narrative art originated right in the midst of the plague outbreak. As Boccaccio states in the introductory part, the idea for his compilation of novellas, which make up the book, was a result of the outbreak of the bubonic plague in the city of Florence in 1348. The made-up stories allowed readers entry into a new universe of mind, a "narraverse," so to speak, which was to the culture of the era what the Metaverse is, arguably, to contemporary culture.

Boccaccio started *The Decameron* in 1348 and finished it in 1375. In the opening part, he explicitly states that he was inspired by the dire living conditions and toll on human life that the plague had brought about, to create an imaginary world that allowed readers a means to escape from actual reality to a more normal reality that people were accustomed to. He described the horrific plight into which the plague had projected people as follows as follows (Bocaccio 1348: vii):

> How many valiant men, how many fair ladies, breakfast with their kinfolk and the same night supped with their ancestors in the next world! The condition of the people was pitiable to behold. They sickened by the thousands daily, and died unattended and without help. Many died in the open street, others dying in their houses, made it known by the stench of their rotting bodies. Consecrated churchyards did not suffice for the burial of the vast multitude of bodies, which were heaped by the hundreds in vast trenches, like goods in a ships hold and covered with a little earth.

The Decameron is a collection of 100 fictional short stories told by a group of seven young women and three young men, who had escaped the Black Death that was afflicting the city of Florence, sheltering in a secluded villa just outside the city. The tales range from stories of love to practical jokes and cautionary tales. It was the first of its kind, as far as can be told, exerting widespread influence, inspiring, for example, Chaucer's

Canterbury Tales. Written in the vernacular of the Florentine language, it is considered by many to be the first masterpiece of narrative fiction in the modern sense of the word. Needless to say, storytelling has existed across time and across cultures, blending elements of truth and fiction. However, it is valid to say that the ancient myths, tales, and fables were not interpreted or perceived in the same way as one would understand fiction in Bocaccio's era or today. One can only imagine what kind of mind-world Boccaccio's stories generated among medieval people, who took to the new narraverse enthusiastically. The world changed irreversibly, as fiction became highly popular, igniting social change across the world. As Umberto Eco (1994: 98) aptly observes:

> To read fiction means to play a game by which we give sense to the immensity of things that happened, are happening, or will happen in the actual world. By reading narrative, we escape the anxiety that attacks us when we try to say something true about the world. This is the consoling function of narrative—the reason people tell stories, and have told stories from the beginning of time.

One could say that *Snow Crash* (1992) is to the Metaverse what *The Decameron* was to the narraverse—a creation of a fictional world shaped by the imagination. Both show a need to escape the real world, at least for a while. The Metaverse is distinguished mainly by the fact that the fiction is ready-made via AI, and does not require a single author. Users can simply log in, perform their activities and customize their avatars, freely creating narrative content, as in a video game.

Epilogue

The American sci-fi author Philip K. Dick, perhaps best known for his story *Do Androids Dream of Electric Sheep* (1968), which was turned into the movie *Blade Runner* (1982), dealt with the theme of simulacra in his appropriately titled 1964 story, *The Simulacra*, which is about a duplicitous government ensconced in what has been called here a simulacrum culture, in which androids in this case are the actual simulacra who live alongside humans. As Matthew Pagan (2023) comments on the work:

> The "simulacra" of The Simulacra are highly functional androids built to resemble human beings. The Ges, the informed upper-class, know full-well that the "husband" of Nicole Thibodeaux, their country's der Alte [president], is non-human; that is, he is a robot simulating the human form: a simulacrum. Many of them work as ersatz engineers, helping to design and build the next planned simulacra. The construction of a der Alte is a task requiring deft art, industrial facilities, and ruthless business cunning. When the country votes every four years to choose between two new husbands for Nicole, both candidates are simulacra.

The fact that Baudrillard (1981) refers to Philip K. Dick in three distinct chapters of *Simulacra and Simulation* suggests that he may been directly influenced by Dick's book, suggesting that even this concept can be traced initially to the narraverse. As Chandra Mukerji (2020) has commented, regarding Boccaccio's *Decameron*, when faced with a dilemma, physical and moral, "people do not need new moral principles to guide them, but rather the means to invent new lives, find inner strength, and laugh off the horror of loss. Boccaccio understood that the plague was a cause and a signal of an old world passing and a new world coming into being." She compares reading Boccaccio's book during the Black Death to binge-watching streaming sites during the covid pandemic, which, like the book, allow viewers to harness the power of fiction to liberate the imagination to seek respite through the narrative format. Boccaccio could have written a treatise on the plague, given his erudition on matters of history; but instead he intuitively understood that narrative fiction was the best way to grasp the meaning of suffering and of an inner sense of the fear that is projected onto human actions. With the *Decameron*, Boccaccio showed the world, of his times and afterward, how surviving something as catastrophic as a pandemic precluded going back to old habits and habitual ways of life. As Mukerji (2020) puts it, "Those who survived found themselves socially untethered and became modern subjects—our forebearers—depending on their wits and imagination to survive."

A guiding principle in this book has been the one used often by McLuhan—namely that when a technology instills itself into everyday life, there is no turning back the clock. The narraverse has given way to the Metaverse, which is a functional simulacrum, a perfect locus for all orders of simulation. It is a prime example of a hyperreal space that creates simulations of reality that are as compelling and convincing as evens in the actual physical world. We may eventually want to live in a video game after all, as *Ready Player One* certainly seemed to imply.

References

Ahn, Sun Joo-Grace, Bostick, Joshua, Ogle, Elise, Nowak, Kristine L., McGillicuddy, Kara T., and Bailenson, Jeremy N. (2016). Experiencing Nature: Embodying Animals in Immersive Virtual Environments Increases Inclusion of Nature in Self and Involvement With Nature. *Journal of Computer-Mediated Communication* 21: 399–419.

Aristotle (c. 320 BCE). *Physics*. Internet Archive, archive.org/stream/in.ernet.dli.2015.183335/2015.183335.Aristotle-The-Physics.

Artaud, Antonin (1938). *Le Théâtre et son double*. Paris: Éditions Gallimard.

Baudrillard, Jean (1981). *Simulacra and Simulations*. New York: Semiotext(e).

Baudrillard, Jean (1988). *The Ecstasy of Communication*. New York: Semiotext(e).

Benjamin, Walter (1934). *Gesammelte Shriften*. Frankfurt: Suhrkamp Verlag.

Boccaccio, Giovanni (1348). *The Decameron*, trans. by Guido Waldman. Oxford: Oxford University Press, 1993.

Brandle, Lars (2023). Motown Enters the Metaverse with Second Life. *Billboard*, https://www.billboard.com/pro/motown-metaverse-second-life-exclusive/.

Broderick, Damien (1982). *The Judas Mandela*. Southlake: Gateway Publishing.

Bühler, Karl (1934). *Die Darstellungsfunktion der Sprache*. Jena: Fischer.

Cover, Rob (2022). Deepfake Culture: The Emergence of Audio-Video Deception as an Object of Social Anxiety and Regulation. *Continuum* 36: 609-621.

Deleuze, Gilles (1968). *Difference and Repetition*. New York: Columbia University Press.

Dick, Phillip K. (1964). *The Simulacra*. New York: Vintage.

Dick, Phillip K. (1968). *Do Androids Dream of Electric Sheep*? New York: Ballantine.

Eco, Umberto (1990). *Travels in Hyperreality*. New York: Harper.

Eco, Umberto (1994). *Six Walks in the Fictional Woods*. Cambridge, MA.: Harvard University Press.

Gorlick, Adam (2011). New Virtual Reality Research. *Stanford Report*, https://news.stanford.edu/news/2011/april/virtual-reality-trees-040811.html.

Hallmon, Dave (2023). Questioning Our Own Simulacrum and Redefining Reality in the Age of Generative AI and Hyper-Realistic Media. *Medium*, https://medium.com/%40DaveHallmon/questioning-our-own-simulacrum-and-redefining-reality-in-the-age-of-generative-ai-and-68762f735b00.

Jameson, Frederic (2007). *The Modernist Papers*. New York: Verso.

Korzybski, Alfred (1933). *Science and Sanity: An Introduction to Non-Aristotelian Systems and General Semantics*. Brooklyn: Institute of General Semantics.

Lanier, Jaron (2017). *Dawn of the New Everything: Encounters with Reality and Virtual Reality*. New York: Henry Holt and Co.

Lanz, Antonio (2023). AI Deepfakes Just Got Better With This Upgrade. *Decrypt*, https://decrypt.co/139118/ai-deepfakes-just-got-better-midjourney-upgrade.

Lavoie, Raymond, Main, Kelley, King, Corey, and King, Danielle (2021). Virtual Experience, Real Consequences: The Potential Negative Emotional Consequences of Virtual Reality Gameplay. *Virtual Reality* 25: 69–81.

Lucretius (c. 99 BCE). *De Rerum Natura*. archive.org/details/onnatureofthing00lucr/page/n5/mode/2up.

McLuhan, Marshall (1964). *Understanding Media: The Extensions of Man*. New York: McGraw Hill.

Mukerji, Chandra (2020). The 'Decameron', or How to Laugh Through a Pandemic. *Public Books*, https://www.publicbooks.org/the-decameron-or-how-to-laugh-through-a-pandemic/.

Nosta, John (2023). The 'Scientific Simulacra': When AI and Hyperreality Collide. *Medium*, https://johnnosta.medium.com/the-scientific-simulacra-when-ai-and-hyperreality-collide-b676eb265045.

Pagan, Matthew A. (2023). The Simulacra by Philip K. Dick. *Medium*, https://www.medium.com/%40mttpgn/the-simulacra-by-philip-k-dick-b4ab7b1d91a9.

Stephenson, Neal (1992). *Snow Crash*. New York: Bantam.

Schwartz, Eric Hal (2023). TikTok Deepfake 'Unreal Keanu Reeves' Spurs Furious Debate. *Voicebot.ai*, can01.safelinks.protection.outlook.com/GetUrlReputation.

Valéry, Paul (1964). *Aesthetics*. New York: Pantheon.

CHAPTER 10

AI and the Future of Pop Culture

Pop culture and entertainment can be dismissed as surface, but it's not. It's the language we all speak, and it's the connection point between people all over the world.
—Bozoma Saint John

Prologue

Throughout this book, examples of AI-generated cultural artifacts, from music to films and podcasts, have been examined in order to assay if pop culture is at a crossroads in terms of its content, forms, and actual makers, morphing more and more into a simulacrum culture (Chapter 9). There is little doubt that the advent of generative AI in the creation of cultural artifacts has broken the historical flow of pop culture in ways that are unprecedented. At the very least, AI has installed itself as a partner with human creators, as audiences seem to be increasingly indifferent as to who or what the source of the artwork is, as long as it moves them in some emotional way. Perhaps this is why there are now even music charts for AI-generated songs, awards for AI-created artworks, prizes for AI-generated fiction, and so on. As the example of "Heart On My Sleeve" (Chapter 1) suggests, as long as people like the music, so be it (no matter who or what created it).

M. Danesi, *AI-Generated Popular Culture*,
https://doi.org/10.1007/978-3-031-54752-2_10

Clearly, we are in the middle of a technological revolution that has significant implications for the future of pop culture as we have known it since its inception in the 1920s. The question to be addressed explicitly in this chapter is the one that has been implicit throughout: Is the shift from real-world human-created pop culture to AI-generated pop culture going to be expanding, and if so, will pop culture, as such, even continue to exist? Pop culture, as a distinct form of culture, with its own artistic and textual categories and genres, crystallized at first as a reaction to the stodgy type of culture in America, where fun and physical enjoyment were downplayed. It spread quickly and broadly throughout the 1920s—an era when it formed a partnership with technology and the business world. The simple reason for its spread was that people sought entertainment massively, which could be delivered through the new technologies. This coalition gave pop culture its identity.

While traditional pop culture has been characterized by constant change, and by many trivial and even inane trends, such as pet rocks and "one-hit" musical wonders, it has nonetheless produced works of lasting value in music, cinema, literature, and other artistic domains, which have shaped the historical flow of the modern world itself. AI-generated culture, on the other hand, is an ad hoc, made-to-order culture, and thus is likely incapable of engendering new movements in the arts or even in the domain of entertainment. However, as discussed throughout, the opposite may also be true—namely, that it might signal the future of the arts themselves, created by a partnership of machines with humans, which may even spur humans themselves to find innovative ways to create new art forms. As Isaac Asimov once put it: "What AI learns, actually, is to be a mirror—to be more like us, in our messiness, our fallibility, our emotions, our humanity" (cited in Dauber 2023).

Among the first to realize the importance of new technologies as adopted by the popular arts to the evolution of society was McLuhan, starting with his 1951 book, *The Mechanical Bride*. McLuhan warned his readers in that book that the mass media to which they were exposed on a daily basis, and the messages that were contained in them, constituted a blessing and a curse at the same time. While they did in fact make information and entertainment available and accessible to larger groups of people, they also tended to engender a feeling of alienation from the makers of artistic content, bringing about a profound change in how people had previously consumed it. Significantly, McLuhan presented his ideas in a style of writing and exposition that mirrored the pastiche *langue* of modern mass culture, using anecdotal and aphoristic style, which went against the traditions of scholarly method, allowing him to tap into the modern mindset and mirror it at the same time, indicating overall that the world was changing radically. He left it up to readers to collate the snapshots into an interpretation of that world on their own terms, perhaps prefiguring how AI systems now allow us to do exactly that with a simple prompt.

Why Pop Culture?

Why is pop culture, *popular*? Is it a culture that stems from the Freudian subconscious and projected onto a social context of artistic populism? Since its beginning, theories to explain the psychological and social functions of pop culture have been abundant. Unlike cultures of the past, pop culture is distinctive in the fact that it meshes art and recreation, engagement and entertainment, seriousness and laughter, in an aesthetic pastiche that has always had great appeal. Alongside fads such as Hula Hoops, pet rocks, Pez dispensers, and the like one finds the music of a George Gershwin and the cinematic art of a Federico Fellini—all of which gain a degree of popularity across all social classes, as they rise intuitively among people and critics alike to the level of high art.

Four major theoretical frameworks have been used widely to explain the psycho-social origins and allure of pop culture—culture industry theory, mythology theory, archetypal theory, and lastly carnival theory. As discussed in various previous chapters, the Frankfurt School scholars were among the first to provide a powerful critique of mass popular culture, seeing it as a product of consumerist society and thus as a commodity culture, where art forms were produced like manufactured commodities —hence their designation of pop culture as a culture industry. True art is not subject to marketplace economics nor can it be reduced to a common denominator in mass tastes. Moreover, technological changes do not affect its aesthetic objectives, and overall meaning (Adorno 1970). But, such critics may have missed the possibility that within pop culture, original and lasting art can indeed crystallize and thrive, as mentioned. The Frankfurt School only saw pop culture as a product of a materialistic-consumerist society, wherein people seek escape from everyday monotony with instant forms of gratification in the world of the arts, virtually in the same way that they seek pleasure in the consumption of products.

Roland Barthes's (1957) mythology theory came forth a few decades after the Frankfurt School approach, becoming a widely-referenced framework for explaining pop culture ever since. Barthes saw in popular entertainments and texts the propensity of modern-day consumerist cultures to engage in the same kinds of themes, narratives, and spectacles of antiquity, carrying them forth in recycled commodified ways, as briefly discussed in Chapter 2. This would explain why, for instance, comic book superheroes have become so popular—they are recycled ancient heroes of mythic and legendary tales. Unlike the original myths, however, which constituted a language for understanding the world in imaginative ways, the modern ones are nothing more that recycled corruptions of these myths. The reconstructed myths, Barthes claimed, manufactured a system of culture that reproduced ancient forms of language and narrativity—and the reason was that, like their ancestors, modern-day people need myth subconsciously to "make things right" in human affairs, at least in the world of fantasy. Because it constitutes a mythological system, Barthes argued that pop culture has had a profound impact on modern-day ethics. In the historical development of ethics, three principal standards of conduct have been proposed as the highest good: happiness or pleasure; duty, virtue, or obligation; and perfection, the fullest harmonious development of human potential. In traditional cultures, these standards were established through religious and philosophical traditions, and expressed in the different forms and rituals associated with folk culture. In pop culture, they are showcased on the mass media, where they are easily manipulated for political or social schemes.

Mythology theory implies that there must be something in the psyche that transcends time, even if in pop culture this occurs in a distorted way, according to Barthes; and this leads us to the doorstep of the Jungian theory of archetypes, which constitutes a third major theory of pop culture. Jung (1959) defined an archetype as an unconscious mental image, pattern, or recurring thought inherited from our ancestors, and passed on and expressed through the cultural channel via recurrent rituals, metaphorical language forms, narratives, and symbolic practices. Archetypes surface in pop culture representations constantly. The Joker, a primary Jungian archetype, surfaces as a villain in the Batman stories; the Shadow, another archetype, is a crime-fighting character in early pulp fiction; and the Hero archetype manifests itself in characters such as Superman and James Bond. Now, for every Hero attempting to overcome some obstacle so to gain some moral objective, there is an older and wiser archetypal character, which Jung called the Sage, who advises and encourages the Hero. Without the Sage, the Hero would never find the strength or wisdom to overcome the hurdles and challenges that stand before them. Sages are everywhere in pop culture narratives, from Yoda in *Star Wars*, Gandalf in *Lord of the Rings*, Dumbledore in *Harry Potter*, Tonto in *The Lone Ranger*, and so on. Needless to say, there are a host of other archetypes in pop culture's representational domain—the point here is that they continue to guide human expressivity, representation, and understanding even in a consumerist world.

The fourth major theory, formulated by Mikhail Bakhtin (1981), introduced the idea that popular spectacles are cathartic, allowing common people to laugh at themselves, to gain enjoyment through music, dance, stories, and other means of popular expression, with no necessary connection to any canons of established aesthetics. Bakhtin claimed that such culture had the same kinds of psycho-social functions as the medieval carnivals. For Bakhtin, people attending a carnival do not merely feel that they are part of an anonymous crowd, as they would be attending, say, an academic lecture or the performance of a symphony in a modern-day concert hall. Rather, they feel part of a communal body, sharing a unique sense of time and space. Through costumes and masks, individuals take on a new identity and, as a consequence, renew themselves psychically in the process. It is through this carnivalesque identity that the grotesque within humans can seek expression through overindulgent eating and laughter, and through unbridled profane expressivity. In such behaviors, people discover who they really are. Carnival theory is based, clearly, on the premise that the human psyche has a contrastive nature—the sacred versus the profane—and that both require expression through language, performance, ritual, and symbolism. This viewpoint is sustained by the fact that many traditional cultures make a distinction between these two in their sets of rituals and in the distribution of their institutions. In Italian traditional culture, for example, the sacred period of Lent is preceded by a carnivalesque day (*Carnevale*), which allows for the release of pent-up urges through song, dance, laughter, costume-wearing, and the like. There is, however, a paradox to be noted. In contemporary pop culture, the line demarcating sacred and profane forms of expression is often nonexistent. Alongside nonsensical movies and banal fads, one finds filmic masterpieces such as *Amadeus* and *Mystic River*, or jazz and rock musical works, such as those by Louis Armstrong and the Beatles, that rise to the level of high art. Perhaps it is this ambiguity that gives pop culture its powerful appeal.

The spread of pop culture has led to the leveling and amalgamation of other cultural forms. Classical music, for example, was seen, before the spread of pop culture, as pertaining to a musically-sophisticated or élite audience, even though such music was meant to appeal to the masses in the late eighteenth and nineteenth centuries. In the modern world, however, classical music is used in commercials, movie soundtracks, and it even has its own fan clubs. Folk traditions also formed a cultural category before the advent pop culture, celebrated at specific times in the calendar year—Christmas, Thanksgiving, Halloween, etc. These are characterized by specific kinds of folk narratives, music, spectacles, and the like, which are ritualistic in nature. But even such categories have undergone "popularization," with Hollywood, television, and other mass media adopting and reshaping them into popular forms and spectacles, having thus redefined the traditions, catapulting them into a cultural admixture, with films, songs, programs, and websites related to the tradition.

Interestingly, Horkheimer and Adorno (1944) actually envisioned beneficial psychological effects of pop culture, calling engagement in the culture as "psychoanalysis in reverse," implying that, like psychoanalytic therapy, pop culture functions as a means to assuage the modern mind that is constantly stressed out by the pressures of materialism and consumerism. Barthes' (1957) mythology theory also implied that pop culture had an important social function, akin to mythic culture in antiquity. For example, the fictional superheroes of pop culture, such as Superman and Wonder Woman, are honest, truthful, physically attractive, strong, and vulnerable, like their ancient counterparts as described in the myths and epic stories such as *The Iliad* and *The Odyssey*; the villains are dishonest, cowardly, physically repulsive, weak, and cunning, also like the scoundrels of mythic yore. Viewing these archetypes in comic books or movies may despoil the ancient stories, Barhes claimed, but at the same time it can be claimed that they provide a new format for recycling them as relevant tales about the human condition.

A Theory of AI-Generated Pop Culture?

So, how does AI-generated culture fit in with theories such as the ones above, if at all? Do we need a new theory? Perhaps not, as can be seen in the fact that video games, for instance, are shaped mythologically, in the

Barthesian sense, given that the same typology of archetypal characters is found in virtually all RPGs. This might explain, in some part, the allure of these games, bolstered by the fact that the mythic narrative is controlled by the players, not by traditions or recycled texts. So, in AI-based video game culture, archetypal and mythological theories are not only applicable, but also reinforced by the fact that agency and control over the narrative is personalized, making the video game even more powerful psychologically.

The same can be said about other areas of AI-generated culture, from music to cinema, given that AI recycles in new ways previous structures of narrativity, visuality, musicality, and so on. Does this imply that AI-generated forms of culture still fall within the "cultural archive," as Lev Manovich (2023) aptly asks—a term referencing anthropologist Wendy James' (1999) notion of culture as the repository of knowledge that manifests itself in everyday interactions? Does AI art indeed fit into the archive? If so, what would it mean for humans to live in a world where a large percentage of stories, musical works, paintings, and so on are not just shaped by nonhuman intelligence, but generated by it? Like the Death of the Author and the Death of the Artist, are we witnessing the Death of Pop Culture (as traditionally understood)? In games like chess and Go, no human can now hope to beat a computer (Chapter 1). What happens when the same situation occurs in art? To cite Manovich (2023):

> We are currently living through early days of a major revolution in communication, media, and culture...For decades we have assumed that AI will not be able to simulate that unique human ability: artistic creativity. Translating between languages, summarizing text, playing chess, and winning Go? Sure. But writing original music, generating detailed photographs of people and objects that do not exist, or making images that perfectly emulate the effects of any media and styles of thousands of well-known artists—and also synthesizing new styles and visual media from combinations of existing ones?...in the past few years, AI scientists have given computers seemingly unique human abilities to create and imagine.

Of course, artists in all domains, from literature and cinema to music and painting, have consistently broken away from the archive, so as to be better able to interpret the world of their times; but these are nonetheless derivatives of the forms in the archive, extended creatively in new forms. Deep neural networks, are trained in an analogous way, as Manovich (2023) goes on to remark—namely "by learning from the historical past and all art created up until now." Of course, maybe AI's combinatorial texts are "creative" in that they reveal new possibilities for art and its methods, reconfiguring them through machine intelligence. In this sense, AI art forms are genuinely new cultural artifacts with contents and styles that humans have never seen before. But then are they creative in any human sense?

AI Creativity

What is creativity, and who (or what) decides what is creative and meaningful? Is AI creativity unique in itself, different from human creativity? If so, how would we be able to recognize it without a new language–perhaps a language that AI may develop on its own? Recall the AI-generated *Portrait of Edmond Belamy*, which sold for almost half a million dollars at auction (Chapter 6). Clearly, it was deemed as creative by those who bought it, being indifferent to the source of the creativity. But then it was our human reaction to the work that was projected onto the evaluation of the machine-produced artwork. One wonders what the AI would have to "say" about its own work, without any form of training or prompting? Would it be capable of doing so, and if so, would we be able to understand its language?

The idea that human art is unique and special, and not reproducible mechanically, is the fact that it is emotional in its inspiration and in its effects on audiences. The rock musician Nicholas Cave, cited in Vaziri (2023), expresses this as follows:

> Songs arise out of suffering, by which I mean they are predicated upon the complex, internal human struggle of creation and, well, as far as I know, algorithms don't feel. Data doesn't suffer. ChatGPT has no inner being, it has been nowhere, it has endured nothing, it has not had the audacity to reach beyond its limitations, and hence it doesn't have the capacity for a shared transcendent experience, as it has no limitations from which to transcend.

Cave injects his own songs with the grief and suffering that tragedy evokes, conveying this in the structure and flow of the melodies, harmonies, and rhythms. His music can be easily described as creative musical art in human terms, given the emotions it evokes. But is simply describing the emotions that it elicits evidence that it is truly creative? As

De Cremer et al. (2023) point out, creativity requires an appropriate language to frame its creativity. Works such as the *Portrait of Edmond Belamy* present a paradox, because they can be described with the same critical language, which means that they are perceived as creative in human terms. As the reliance on generative AI becomes more and more embedded in the production of culture, and given the ability of AI to produce large quantities of artworks in minutes, we will need a new critical apparatus for discussing art as a mass-production phenomenon. So, in this AI environment, would we even be able to recognize art forms as novel and authentic, rather than aleatory creations based on algorithmic processes? The ancient stories, such as the *Iliad* and the *Odyssey* were composed not as "out-of-the-blue" creations, but as narrative epic poems within the traditions of historical storytelling of the era. They also had a human purpose—to explain character and human-perceived reality in narrative ways that were understandable to everyone via verisimilitude. One wonders if this use of literary creativity, which is now part of the cultural archive, will ever be matched by AI creativity.

In a relevant research study, Rhiannon Williams (2023) found that it is difficult to pin down what AI creativity is, without human linguistic assessments, and therefore: "It could just be that AIs can pass creativity tests, not that they're actually creative in the way we understand." In other words, we really have no access to what true, autonomous machine creativity would be. We can only describe the algorithm involved—in human terms. In fact, without language, it would be a moot point to even contemplate what creativity is. Language is the cognitive apparatus for *thinking* about texts such as songs and pantings, impelling us to consciously reflect on: (a) how the texts mean what they mean, (b) how they allow us to interpret feelings rather than just refer to them, and (c) what aspect of reality they bring to consciousness that may not have been obvious beforehand. It is not clear at present what to make of AI creativity by using the same language to assess it, since the creativity in this case is realized algorithmically. Only if AI can somehow describe its own creativity, with its own algorithmic-mathematical language, will we be able to make a meaningful assessment about what AI creativity might entail in comparative terms—that is, by comparing traditional accounts of human creativity with the machine's mathematical accounts.

Where's the Beef?

"Where's the beef?" is a catchphrase that was introduced as a slogan by the fast food chain Wendy's in 1984 (Cross 2023). Since then, it has become an all-purpose phrase for questioning the substance of an idea. By semiotic extension, "Where's the beef" can be reformulated as "Where's the meaning?" But then, what is meaning? As Ogden and Richards showed in their masterful 1923 work, titled appropriately *The Meaning of Meaning*, there are many meanings of the word "meaning" in English, making it impossible to pin down what is intended with the word "meaning". So, within semiotics a more precise lexicon for discussing "meaning" was developed, including a basic distinction between *denotation* and *connotation*—which can actually be traced back to the medieval Scholastics. To see what this might imply, consider how we interpret a sitcom , as representing a situation that occurs in real life. This is the denotative meaning that we glean from the sitcom; it is the first level of analysis, based on what we actually see going on in the sitcom. But our interpretation of the sitcom hardly ends there. It is when we connect the sitcom to events, changes, and ideas to an extant social code (Chapter 5) that it transcends its denotative meaning. This is the second connotative level of meaning, also called he "coded" level (Barthes 1957). It is level where we bring emotional and cultural associations to bear on the interpretation of the sitcom. Of course, a chatbot can access that code, if trained on it, and paste together a sitcom with similar connotative meanings. But it has no (human) consciousness of the connotations; it just collates them according to specific prompts. The sitcom's "beef" is in the human mind, not in the algorithm, as we saw with *Nothing, Forever* (Chapter 5).

The Wendy's slogan was in perfect harmony with the societal mood of the 1980s—an era defined by a desire for authenticity and substance. The advertising narratives woven around the catchphrase created "a profound emotional connection" (Cross 2023) to meaning-making in an increasingly nihilistic society, enabling audiences to seek meaning even in a simple hamburger, evoking the objectives of the pop artists. Now, recall the AI-generated beer commercial "Synthetic Summer" (Chapter 3). At a denotative level, it can easily be perceived as depicting real people and a realistic location. But as the video progresses, the AI either did not grasp the laws of physics, or else became overly creative (hallucinatory) after discovering what fire is (from the Internet). Its deepfake qualities become increasingly apparent as we look at it more closely, including the fact that the beer can in the commercial has an artificial quality, which makes it an abstraction rather than an actual object. In sum, the AI seems to not have understood the human connotations involved in the scene, as it recreated it algorithmically. The commercial leaves us wondering where the beef is, even though denotatively it certainly appears to be a normal commercial for a beer product.

Again, language is the key to all this, given that it is the "operating system" of human culture, to utilize a computer metaphor. Without it, even a simple explanation about what the commercial above denotes, would be impossible. While AI's new mastery of language means it can allow it to "hack" the human "operating system," via training and prompting, it clearly does not understand it in human terms. To extend the beef metaphor, AI could easily create the meat (the actual texts), but then would it be able to distinguish its taste from other tastes? As Ben Dickson (2018) has aptly commented on this whole line of discussion:

> But mastering human language requires much more than replicating human-like voices or producing well-formed sentences. It requires commonsense, understanding of context and creativity, none of which current AI trends possess...AI is a technology for augmenting humans and can help speed or ease tasks that involve the use of human language. But still lacks the commonsense and abstract problem-solving capabilities that would enable it to fully automate disciplines that require mastering of human language.

Six Characters in Search of an Author

The 1921 play, *Six Characters in Search of an Author*, by Luigi Pirandello, represents one of the first dramas of so-called meta-literature—a play about where the meaning of a work liesin an author, in an audience, or elsewhere. An acting company prepares to rehearse a play. As the rehearsal is about to start, the group is unexpectedly interrupted by the arrival of six uninvited eccentric actors, who ask for an author to write them into existence, explaining that they were abandoned by their own author, but still want to have their story completed. The director initially believes that the people had escaped from an asylum. However, his artistic sensibility became piqued by the "incomplete story" that they tell him, and so he agrees that it should be realized theatrically. They are thus given existence textually and representationally.

The play raises questions that are more resonant today than they were when it was first performed. A human author can write a story for characters, since these are imaginary figments constructed to represent real people, giving them the fullness of literary life. Now, would an AI system be able to write a story for these characters that would make sense in a similar thematic way? It could certainly write up a hypothetical story, but would it match Pirandello's marvelous treatment of the difference between reality and illusion? As Oliver Tearle (2021) notes:

> To whom does a story or a scene 'belong'? The author of the play? The director who brings it to life on the stage? The actors who 'interpret' the part? Or, if the play is based on real-life events, does it truly belong to the people who lived through the events the play dramatises?

It is certainly extraordinary to note that AI has become an Author, on its own, as discussed in Chapter 2. It clearly has the ability to "write plays," as evidenced by the AI script of a play called *Ten Strangers in a Room*, which was performed at the University of Wollongong in Australia in 2023—a play completely generated by a chatbot. The play was about a group of strangers discussing their lives and interests during a house party. The human director lamented, however, that it: "was bad drama, like a bad play. The language is quite generic and cliched, the characters are superficial and underdeveloped and all of the drama is sort of told to you as the audience" (cited in Huntsdale 2023). One of the student actors commented that, while the script made sense (denotatively), it was lacking in depth: "AI is never going to be able to reflect genuine human emotions, she said. It can create a caricature where it will tell you exactly how someone is feeling and you lose those nuances, so we are telling, not showing" (cited in Huntsdale 2023).

So, what makes a work of art meaningful in human terms? One factor is motivation, a psychic state that propels people to engage in goal-directed behavior. It is unlikely that an AI system will ever become motivated to compose scripts, music, art, etc. *on its own*, without algorithmic training. A second is inspiration. No matter how nebulous this term is, we use it commonly to explain why a play (or any other creative text) comes about in the first place. With an AI work such as the *Portrait of Edmond Belamy*, the "inspiration" is certainly not in the algorithm, which had to be trained by a human. So, the source of inspiration is not in the machine, but in the human programmer who transfers it to the machine via algorithmic training. A third factor is transcendence, in the sense of a work retaining its aesthetic value in eras subsequent to when it was created. Pirandello's play, written in 1921, has meaning and relevance today and in all likelihood tomorrow, transcending the time in which it was written. The AI play *Ten Strangers in a Room* will be nothing more than an oddity in the future.

Sentient AI might of course produce its own (untrained) authentic works of art. But if this were to happen, how would we know, asks Elizabeth Finkel (2023)? We would have to enter into the"mind" of the AI, and AI would have to enter the human mind as well, without training or prompting. Perhaps as Ray Kurzweil (2005) has claimed, there will inevitably come a moment in time when AI will have progressed to a point that it will autonomously outperform human intelligence and become authentically creative on its own. At that point, the AI must surely be considered conscious, whatever that means. Does it mean that machines and humans will become rivals of each other? This would then mean that AI would no longer generate a synthetic culture, but a truly original one. At that point, would human culture continue along its historical path, integrate with AI culture, or morph into something completely different (Settembre 2020)?

Epilogue

Any foray through AI-generated culture will beg a fundamental question related to the human condition: Are we no longer evolving through biology, but through our technology? As McLuhan (1968: 65) aptly remarked, humans live in an "environment that transfers the evolutionary process from biology to technology." Certainly, as discussed throughout this book, pop culture was made possible by the creative uses of new technologies, from sound recordings to cinema and television, but not by the technologies in themselves, which at the time were lacking AI. It is little wonder that one of the concerns of pop culture itself has actually been the possibility of our technologies coming to life, textualized narratively in many works of sci-fi (Chapter 1).

As McLuhan maintained (1964), each major change in technology invariably brings about an evolutionary shift in human consciousness and spurs on innovations in cultural life. To support this argument, McLuhan used the example of the invention of writing, which required the creation of physical tools to be realized. When movable type technology extended the reach of written materials, a new world order fomented that he called the “Gutenberg Galaxy,” after Johannes Gutenberg the German printer who perfected this technology in Europe in the mid-1400s. Now, one can claim that all this is simply an interesting argument, which cannot be proved or disproved. There could, in fact, be other ways of explaining how the world had changed after the invention of writing. That said, there is little doubt that it strikes an unconscious chord, providing a conceptual template through which we can understand social evolution and its connection to technology. If writing had never been invented, the world would be drastically different.

Writing constituted the first paradigm shift in cognition and consciousness; AI technologies may constitute the second such shift. The radical part of the AI paradigm is the ability of AI to perform human creative tasks, even if they are, at present, “interpretant-less,” floating exclusively to the human user. Clearly, the advent of AI-generated culture raises many questions, which can only be discussed at present by humans, or prompted by humans to a chatbot. As in any major change brought about by a new technology, we are clearly in the throes of a social transformation, which might actually be highly beneficial, since it is inducing a reappraisal of our notions and concepts about art, as has been done throughout this book. As De Villamil (2023) puts it:

> [AI] will reevaluate the past, transforming our perception of art, music, and film. The algorithms will pore through archives, unearthing hidden gems of creativity that have long been forgotten. Artists of the past will be resurrected, reimagined, and presented to a whole new generation, bridging the gap between eras and inspiring fresh waves of artistic expression. Through AI, pop culture will become a captivating blend of human creativity and machine ingenuity. The line between creator and creation will blur as AI algorithms generate content that resonates with us on a deeply personal level. Our playlists, reading lists, and viewing preferences will be tailored with an eerie accuracy, as AI uncovers our deepest desires and anticipates our next cultural fix.

To conclude this foray into the implications of AI-generated pop culture, it is fitting to turn to sci-fi one last time, a genre that has guided the present journey all along—and specifically to the remarkable film *Blade Runner* (1982). The key question that the film asks is what it means to be human, and, as a corollary, if something that is not an organic being can be "alive." Interestingly, in the movie, it is the non-organic replicants that ask that very question, not the humans. Replicants are implanted with memories to make them believe that they are alive. Throughout the film, the replicants show emotions even more than humans, so: Does it matter that the replicants were created artificially instead of being born through a human body? The answer is suggested in the final scene of the movie. The protagonist, Deckard, runs off with the replicant Rachel, with whom he had apparently fallen in love, as did the replicant with him. Both know that Rachel is a replicant, but they do not let this fact get in the way of how they feel toward one another. The implication is that love may be an emotion that is independent of any system, organic or non-organic, existing immanently in the world as a beneficial force that seeks expression in many ways. Interestingly, the movie interweaves a number of hints that Deckard might unknowingly be a replicant himself. In the end, it does not matter in any practical sense, since life goes on, naturally and artificially.

In Sumerian and Babylonian myths, there were accounts of the creation of life through the animation of clay (Watson 1990: 221). The ancient Romans were also fascinated by automata as artificial life forms. By the time of Mary Shelley's *Frankenstein*, the idea that robots could be brought to life was given a fictional form suited to the popular imagination at the time. Since then, the quest to animate machines has been relentless, in science and in sci-fi. Movie robots have all the attributes of gods, retrieving the mythic instinct in all of us. So, as Watson (1990: 228) notes, any sense of unease with this type of technology is not so much "a premonition of disaster as nothing more than simple incredulity."

References

Adorno, Theodor (1970). *Aesthetic Theory*. London: Bloomsbury.

Bakhtin, Mikhail (1981). *The Dialogic Imagination: Four Essays*. Austin: University of Texas Press.

Barthes, Roland (1957). *Mythologies*. Paris: Seuil.

Cross, David (2023). Where's the Beef? Catchphrase Case Study. *Medium*, https://medium.com/%40david_6844/wheres-the-beef-catchphrase-case-study-67f7ea2e5978.

Dauber, Jeremy (2023). What Isaac Asimov Can Teach Us about AI. *The Atlantic*, https://www.theatlantic.com/books/archive/2023/03/ai-robot-novels-isaac-asimov-microsoft-chatbot/673265/.

De Cremer, David, Bianzino, Nicola Morini, and Falk, Ben (2023). How AI Could Disrupt Creative Work. *Harvard Business Review*, https://hbr.org/2023/04/how-generative-ai-could-disrupt-creative-work.

DeVillamil, Eugene (2023). Sex, Drugs, Rock-n-Roll, and AI: The Futuristic Transformation of Pop Culture. *Medium*, medium.com/aimonks/sex-drugs-rock-n-roll-and-ai-the-futuristic-transformation-of-pop-culture-51fe4364c630.

Dickson, Ben (2018). A Reality Check on AI's grasp of Human Language. *TechTalks*, https://bdtechtalks.com/2018/10/22/ai-deep-learning-human-language/.

Finkel, Elizabeth (2023). If AI Becomes Conscious, How Will We Know? *Science*, https://www.science.org/content/article/if-ai-becomes-conscious-how-will-we-know.

Horkheimer, Max and Adorno, Theodor (1944). *Dialectic of Enlightenment: Philosophical Fragments*. Stanford: Stanford University Press.

Huntsdale, Justin (2023). Theatre Play Written by Artificial Intelligence Website Chatgpt Performed at University of Wollongong. *ABC News*, https://www.abc.net.au/news/2023-06-10/play-written-by-chatgpt-ai-performed-at-university-of-wollongong/.

James, Wendy (1999). *The Listening Ebony: Moral Knowledge, Religion, and Power Among the Uduk of Sudan*. Oxford: Oxford University Press.

Jung, Carl (1959). *The Archetypes and the Collective Unconscious*. Princeton: Princeton University Press.

Kurzweil, Ray (2005). *The Singularity Is Near*. Harmondsworth: Penguin.

Manovich, Lev (2023). The AI Brain in the Cultural Archive. *Museum of Modern Art Magazine*, https://www.moma.org/magazine/articles/927.

McLuhan, Marshall (1951). *The Mechanical Bride: Folklore of Industrial Man*. New York: Vanguard.

McLuhan, Marshall (1964). *Understanding Media: The Extensions of Man*. Cambridge: MIT Press.

McLuhan, Marshall (1968). *Through the Vanishing Point*. New York: Harper & Row.

Ogden, Charles K. and Richards, I. A. (1923). *The Meaning of Meaning*. New York: Harcourt, Brace and World.

Settembre, Amelia (2020). Living in a Base Reality—Are We in a Simulation? *Medium*, https://medium.com/%40amesett/living-in-a-base-reality-are-we-in-a-simulation-1b819e46aaed.

Tearle, Oliver (2021). A Summary and Analysis of Luigi Pirandello's Six Characters in Search of an Author. *Interesting Literature*, https://interestingliterature.com/2021/02/pirandello-six-characters-in-search-of-an-author-summary-analysis/.

Vaziri, Aidin (2023). Nick Cave Slams AI-Generated Song by S.F. Company, Calls it 'Grotesque Mockery'. *Datebook*, datebook.sfchronicle.com/music/nick-cave-slams-ai-generated-song-by-s-f-company-calls-it-grotesque-mockery.

Watson, Lyall (1990). *The Nature of Things*. London: Houghton and Stoughton.

Williams, Rhiannon (2023). AI Just Beat a Human Test for Creativity. What Does That Even Mean? *MIT Technology Review*, https://www.technologyreview.com/2023/09/14/1079465/ai-just-beat-a-human-test-for-creativity-what-does-that-even-mean/.

Index

A

abstract art, 9, 107

Action Comics, 37

Adorno, Theodor W., 76, 77, 83, 138, 139, 177, 180

ads, 78, 128, 130, 132–138, 140

advertising, 4, 8, 107, 108, 128–140, 184

African Americans, 70, 71

AI Age, 12, 40, 46, 73, 76, 92, 98, 117, 127, 132

A.I. Artificial Intelligence, 8

AI Ashley, 87–90, 92

AI SpongeBob, 95

AIVA (Artificial Intelligence Virtual Artist), 3, 68, 78

al-Ghaili, Hashem, 56

algorithms, 1, 3, 10, 18, 29, 32, 34, 37, 42, 60, 61, 68, 72, 90, 96, 102, 105, 106, 110, 112, 113, 115, 119–121, 133, 143, 150, 153, 182, 183, 186, 188

Allegory of the Cave, 151

All That Jazz, 70, 80

AllttA, 16, 73, 74

AlphaGo, 18

Amper Music, 3, 68

androids, 9, 80, 81, 94, 171, 172

archetypes, 32, 178

Aristotle, 15, 35, 100, 168

art criticism, 119

artificial intelligence (AI), 1–5, 7, 8, 10–12, 14–19, 23–26, 28–38, 40–42, 45–47, 49–62, 67, 68, 71–79, 82, 83, 87–90, 92–97, 99, 100, 102, 105, 106, 109–117, 119–124, 127–129, 131–140, 143, 144, 146, 148–150, 152–155, 159–164, 166, 168, 175–177, 180–188

artworks, 9, 10, 42, 106, 107, 111, 113, 116, 120, 121, 166, 168, 175, 182, 183

Asimov, Isaac, 5, 6, 8, 176

Assassin's Creed, 148

aura, 42, 106, 117–119, 168

M. Danesi, *AI-Generated Popular Culture*,
https://doi.org/10.1007/978-3-031-54752-2

author, 23, 24, 26–30, 32–34, 42, 56, 59, 79, 80, 112, 115, 116, 155, 169, 171, 185, 186
authorship, 24, 29, 41, 42, 111, 115, 116, 162

B

Bakhtin, Mikhail, 27, 30, 33, 41, 59, 179
Barbie, 132
Barthes, Roland, 24, 27, 29, 31, 36, 38, 114–116, 130, 178, 180
Baudrillard, Jean, 15, 16, 41, 47, 137, 138, 147, 159–163, 168, 172
Beatles, 72, 123, 179
Beethoven, Ludwig van, 3, 11, 75, 77, 119
Benjamin, Walter, 42, 106, 117, 136, 168
Berry, Chuck, 83
Birth of a Nation, 46
Black Death, 170, 172
Blade Runner, 6–8, 17, 62, 171, 188
Boccaccio, Giovanni, 15, 24, 170, 172
books (print), 37
Bostrom, Nick, 144, 151, 152
British Pears Soap, 130
broadcasting, 10, 89–91, 93, 94
Bühler, Karl, 168
Bunyan, John, 27
Butler, Samuel, 5, 7, 25

C

Čapek, Karel, 6
carnival theory, 177, 179
Cave, Nicholas, 182
celebrity, 2, 24, 46, 70, 74, 107, 109, 110, 128–130, 136, 155, 161, 163, 166, 168
Cent mille milliards de poèmes, 31, 32
Charleston, the (dance), 69
Chasing Waterfalls, 79
chatbot, 26, 31, 53–56, 94, 99, 102, 139, 184, 186, 188
ChatGPT, 17, 25, 26, 34, 46, 54, 55, 133, 134, 149, 182
Chaucer, Geoffrey, 15, 24, 170
Chicago (movie), 70, 80
cinema, 6, 45–51, 53, 57–59, 61, 89, 100, 176, 181, 187
Cinema Paradiso, 57
classical music, 3, 68–70, 75–79, 89, 180
Coca-Cola, 109, 133, 137, 140
code (in semiotics), 13, 14, 121, 184
comics, 37–39, 100, 107, 108, 154, 178, 180
commercials, 4, 30, 52, 54, 107–109, 128–132, 138, 163, 180, 184, 185
commodification, 94, 138, 139
connotation, 184
consumerism, 130, 132, 138, 141, 180
context, 14, 15, 17, 28, 76, 98, 113, 115, 118–120, 122, 177, 185
counterculture, 68, 71, 107, 128
crazes, 69, 127, 133, 146
creativity, 2–5, 9, 12, 14, 18, 19, 32, 34, 35, 37, 38, 40, 41, 47, 50, 51, 53, 56, 62, 75, 79, 91, 102, 110, 113, 118–121, 133, 137, 181–183, 185, 188
Crow, The, 60
culture jamming, 134, 135

D

Dadaism, 36
Daddy's Car, 72
DALL-E, 111, 114–116, 133
Dante, 24, 34

Day a Computer Writes a Novel, The, 23, 25, 28, 30, 31, 37
Death of History, 41
Death of the Artist, 114, 181
Death of the Author, 27, 29, 36, 41, 114, 181
Decameron, The, 170–172
Deep Blue, 1
deepfake, 4, 11, 14, 17, 52–54, 74, 90, 93, 95, 99, 110, 131, 133, 135, 136, 139, 161, 166, 168, 169, 184
deep neural network, 39, 181
Deleuze, Gilles, 16, 162
DeMille, Cecil B., 58
Denise, 90
denotation, 184
De Rerum Natura, 161
Derrida, Jacques, 26
'desert of the real', 16, 147, 162, 163
Dick, Philip K., 7, 8, 171, 172
dime novel, 26, 33, 69, 100
disco, 70
Disney, Walt, 52
disparition, La, 30
Divine Comedy, vi
Do Androids Dream of Electric Sheep?, 7, 8, 171
Dominoes, 74
Drake, 11, 14, 67, 166
Dungeons & Dragons, 145

E
Eco, Umberto, 24, 26, 28, 48, 59, 153, 162, 171
Edison, Thomas, 46, 48
Edmond de Bellamy, 110, 111
Electronic Age, 12
Eliot, George, 7, 8
Eliza, 17, 18
embodiment, 5
Emilie Arnoux, 109
epochs, 12, 148
Erewhon, 5, 7, 25
existentialism, 56
eXistenZ, 150
Ex Machina, 6, 61, 62

F
fads, 130, 177, 179
fantasia, 14–17, 51, 118, 119
fiction, 15, 25–28, 33, 34, 37, 39, 41, 59, 89, 93, 114, 161, 169–172, 175
filmmaking, 2, 10, 45, 49–51, 53, 55–58
Finnegan's Wake, 28
flapper, 71
floating interpretant, 60, 79, 114, 115
floating signifier, 59
FN Meka, 10
Focardi, Giovanni, 130
Foucault, Michel, 29, 31
Frankenstein, 5–7, 62, 189
Frankenstein complex, 5
Frankfurt School, 68, 82, 92, 94, 102, 131, 138, 139, 177
Frost, The, 2, 3, 45, 50–52, 60, 61
Fusion, 79
Futuri Media, 87

G
Gaddis, William, 73
gaming culture, 143, 166
Gargantua and Pantagruel, 27
Generative AI, 3, 4, 11, 58, 79
Given Again, 55
global village, 12
Goodwin, Ross, 24, 42
Griffith, D.W., 46
Gutenberg Galaxy, 188

H

Hamilton, Richard, 108
Harry Potter, 4, 52, 53, 131
Harry Potter by Balenciaga, 4
Heart on My Sleeve, 11, 13, 14, 67, 68, 71, 73, 82, 166, 175
Heilig, Morton, 164
hip-hop, 9, 11, 13, 16, 17, 71, 79
history, 2, 3, 8, 12, 24, 34, 39–41, 58, 73, 88, 98–100, 102, 119, 121, 141, 154, 172
Hogan's Alley, 37
homo ludens, 146
Horkheimer, Max, 76, 138, 180
Huizinga, Johan, 146, 147
hyperreality, 16, 17, 47, 48, 52, 136, 138, 140, 149, 162, 163, 168
hyperreal marketing, 129, 136–138, 140
hyperstition, 138, 152

I

I Am Code, 37
iconicity, 149
Iliad, 6, 180, 183
Illiac Suite, 75, 76, 78
Impressions of Theophrastus Such, 7
inspiration, 26, 33, 51, 71, 83, 132, 186, 187
interpretant, 13, 14, 17, 28, 29, 32, 40, 51, 59, 60, 72, 78, 79, 97, 106, 116

J

Jay-Z, 17, 73, 74
jazz, 9, 68–71, 73, 82, 83, 92, 123, 179
Jeopardy, 2
Joe Rogan AI Experience, The, 93
Johns, Jasper, 107
Joyce, James, 28

K

Kasparov, Garry, 1
Kerouac, Jack, 23, 24, 27, 41, 42
Kiss, The, 46
Kizuna, 159, 163
Korzybski, Alfred, 9, 117, 160
Kubrick, Stanley, 55
Kuleshov Effect, 48
Kurzweil, Ray, 187

L

Langer, Susanne, 119
Lang, Fritz, vii
langue, 116, 136, 176
Lanier, Jaron, 56, 62, 164
Last Stand, 56
Lee, Angus, 79
Lee, Spike, 70
Lévi-Strauss, Claude, 59
lipogram, 30
literacy, 57, 91, 100
literature, 15, 24, 25, 28, 30, 32–35, 38, 114, 160, 163, 176, 181
Lotman, Juri, 121, 122
Lucas, George, vii
Lucretius, 161

M

machine learning, 1, 3, 25, 77, 106
Marcuse, Herbert, 76, 131
marketing, 107, 128–130, 136, 139, 140
Marshall, Glenn, 60
mass culture, 91, 107, 139, 177
mass media, 76, 88, 89, 98–102, 176, 178
mathematics, 15, 78
Matrix, The, 6, 16, 47, 54, 56, 136, 150–152, 155, 162
Maybellene, 83
Mazes and Monsters, 146

McLuhan, Marshall, 3, 12, 24, 45, 46, 61, 73, 75, 88, 89, 100, 101, 172, 176, 187
mediated social code, 97, 98, 100, 102
metaphor, 36, 55, 62, 107, 137, 163, 185
Metaverse, 161, 169–172
Metropolis, vii
Midjourney, 111, 112, 116, 117, 134, 167
Mindscape, 25
Miquela, 74
Mona Lisa, 166, 167
motivation, 35, 71, 76, 95, 186
movies, 3, 6–8, 16, 17, 45–61, 69, 70, 74, 78, 80, 131, 132, 136, 146, 150, 152, 155, 160, 162, 164, 165, 167, 171, 179, 180, 189
movie theater, 46–49, 57
MSCHF, 127–129, 131, 133, 136–141
MuseNet, 10
music, 3, 4, 9–11, 14, 17, 53, 58, 60, 67–80, 82, 83, 90, 92, 102, 107, 123, 131, 134, 150, 155, 169, 175–177, 179–182, 186, 188
myth, 71, 75, 178
mythology, 56, 138, 145, 177, 178, 180

N
narrativity, 26, 28, 39, 144, 152, 153, 178, 181
narraverse, 170–172
neomania, 130
NewsGPT, 95, 96
Nike, 133
No Man's Sky, 143, 146, 148, 150, 152, 153
non-player characters (NPCs), 144, 149
Noonoouri, 74
Nothing, Forever, 95, 98, 102, 184
novels, 2, 3, 6–8, 13–15, 18, 23–28, 30–36, 38, 41, 42, 46, 48, 51, 53, 68, 73, 92, 113, 116, 136, 153, 155, 164, 169, 183

O
Odyssey, 183
On the Road, 23, 24, 42
OpenAI, 10, 93
opera, 79, 80, 100, 112, 123
orality, 88
or inspiration, 76
Oulipo, 30, 35
Outcault, Richard Felton, 37

P
Peirce, Charles, 116, 149
penny arcade, 145
Pepperoni Hug Spot, 134
Perec, Georges, 30
pet rock, 128, 129, 176, 177
Picasso, Pablo, 113, 120
Pilgrim's Progress, 27
pinball machine, 145, 146
Pirandello, Luigi, 185–187
Plato, 79, 151, 152
player piano, 72, 73
podcast, 68, 88, 93, 94, 175
poetry, 8, 14, 31, 34–38, 41
Policeman's Beard, The, 25, 28
pop art, 106–110, 123, 124, 129, 130, 133
popular culture, 33, 57, 123, 145
Postman, Neil, 6, 19, 93, 99
postmodernity, 147
Print Age, 12, 24
Propp, Vladimir, 26, 153

Proteus Effect, 154
pulp fiction, 26, 28, 38, 69, 100, 178
punk, 70, 107

Q

Queneau, Raymond, 31, 35

R

Rabelais, François, 27
Racter, 25
radio, 2, 10, 38, 47, 70, 75, 82, 87–94, 97, 98, 100, 129, 131
RadioGPT, 87, 90
randomness, 36, 146
Rapper's Delight, 71
Rauschenberg, Robert, 107, 129
reader, 24, 27–31, 34–36, 39, 41, 114–116, 176, 177
Ready Player One, 152, 163, 173
replica, 87
replicant, 7, 8, 17, 189
representation (in semiotics), 13, 14, 121, 122, 149
reproduction, 42, 73, 106, 117, 160, 168
Road, The, 23, 24, 27–29, 32, 41
Roblox, 137, 170
robots, 5–7, 17, 62, 73, 74, 80–82, 89–91, 172, 189
rock and roll, 9, 70, 71, 123
role-playing game (RPG), 144–149, 152–155, 181
Runaround, 6
Runnin' Wild, 69

S

Samuel, Arthur, 1
Sarnoff, David, 91
Saussure, Ferdinand de, 99, 116
Savages, 16, 73
Scary Beauty, 80
science fiction, 5, 7, 8, 49, 62, 164
Scott, Ridley, 7
Searle, John, 29
Seinfeld, 95, 96
semiosis, 12–14, 16, 17, 121, 122
semiosphere, 121, 122
semiotics, 2, 13, 14, 28, 72, 106, 121, 122, 149, 168
Sensorama, 164
Sgt. Pepper's Lonely Hearts Club Band, 123
Shelley, Mary, 5–7, 62, 189
simulacrum, 16, 52, 74, 140, 147, 160, 162, 164, 165, 169, 170, 172
simulacrum culture, 159–161, 163, 164, 166, 168, 169, 171, 175
simulacrum effect, 15, 52, 53, 68, 147, 150, 166
simulation, 6, 38, 52, 54, 68, 71, 137, 138, 140, 141, 144, 147, 149–152, 156, 160–165, 172
simulation hypothesis, 6, 144, 148, 150, 154
Snow White and the Seven Dwarfs, 52
Sousa, John Philip, 73
Spielberg, Steven, 8, 55
Stable Diffusion, 113, 114, 135
Star Wars, 160, 168, 179
story-telling, 26, 27, 154, 171, 183
streaming, 49, 57, 67, 79, 93, 95, 172
STYNGR, 169
subvertisement, 134, 141
Sugar Hill Gang, 71
Summer of Sam, 70
Sunspring, 49, 51, 53
Super Mario Bros, 153
surrealism, 56, 60, 61
symphony, 3, 10, 77, 79, 179
Synthetic Summer, 54, 55, 184

T

technology, 2, 3, 6–8, 10–12, 16, 18, 19, 34, 40, 45–47, 49, 51, 57, 62, 72, 73, 75, 78, 81, 82, 87–96, 99, 100, 123, 124, 129, 131, 133, 139, 140, 144, 145, 147, 148, 154, 159–161, 164–166, 169, 172, 176, 185, 187–189
technopoly, 6, 92
television, 10, 38, 48, 49, 61, 89, 94, 95, 98–101, 129, 145, 146, 163
Ten Strangers in a Room, 186, 187
Terminator, The, 56, 160
text (in semiotics), 13, 28
Théâtre d'Opéra Spatial, 112, 117, 118
Theory of Mind, 59–62
The Weeknd, 11, 14, 67, 166
Three Laws of Robotics, 6
Thurber, James, 30, 31
time-binding, 9, 14, 24, 40, 117
Tin Pan Alley, 69
Tornatore, Giuseppe, 57
Turing Test, 5, 8, 62, 75
2001: A Space Odyssey, 55, 148
typographic brain, 88, 89
Tzara, Tristan, 36

U

Uexküll, Jakob von, 5
Unreal Keanu Reeves, 167, 168

V

Velvet Underground, 107, 123
Vico, Giambattista, 14, 35, 118, 119
video game, 10, 25, 78, 128, 141, 143–145, 149–152, 154–156, 160–163, 165, 166, 169, 170, 173, 180, 181
Virtual Reality (VR), 152, 159–166, 169
Voight-Kampff Effect, 9–11, 17
Voight-Kampff Test, 8, 9, 17, 34, 45, 74

W

Wachowski, Andy, 47
WALL-E, vii
Warhol, Andy, 107, 109, 110, 123, 129, 130, 141
Watson, Lyall, 2, 51, 189
Weizenbaum, Joseph, 18
White, Hayden, 39
Wiener, Norbert, 81
Will Smith deepfake video, 135, 168
Wise Little Squirrel, The, 25, 26
Wonderful O, The, 30
writing, 3, 12, 23, 24, 28, 32–36, 39, 40, 42, 45, 47, 49, 53, 58, 76, 83, 87, 94, 96, 100, 121, 176, 181, 188

X

Xiaole, 80
Xinhua, 90

Y

yellow kid, 37
You Dirty Boy!, 130

GPSR Compliance
The European Union's (EU) General Product Safety Regulation (GPSR) is a set of rules that requires consumer products to be safe and our obligations to ensure this.

If you have any concerns about our products, you can contact us on

ProductSafety@springernature.com

In case Publisher is established outside the EU, the EU authorized representative is:

Springer Nature Customer Service Center GmbH
Europaplatz 3
69115 Heidelberg, Germany

www.ingramcontent.com/pod-product-compliance
Lightning Source LLC
Chambersburg PA
CBHW072307300426
44234CB00004B/326
* 9 7 8 3 0 3 1 5 4 7 5 1 5 *